THE SCIENCE OF
STRESS

THE SCIENCE OF
STRESS

LIVING UNDER PRESSURE

GREGORY L. FRICCHIONE, ANA IVKOVIC, ALBERT S. YEUNG

THE UNIVERSITY OF CHICAGO PRESS

Chicago and London

The University of Chicago Press, Chicago 60637

Text © Gregory L. Fricchione, Ana Ivkovic, and
Albert S. Yeung 2016

Design and layout © The Ivy Press Limited 2016

Consultant Editor: Gregory L. Fricchione

Printed in China

26 25 24 23 22 21 20 19 18 17 16   1 2 3 4 5

ISBN-13: 978-0-226-33869-9 (cloth)
ISBN-13: 978-0-226-33872-9 (e-book)
DOI: 10.7208/chicago/9780226338729.001.0001

Library of Congress Cataloging-in-Publication Data

Names: Fricchione, Gregory, author. | Ivkovic, Ana, author. |
    Yeung, Albert, 1955– author.
Title: The science of stress : living under pressure / Gregory
    Fricchione, Ana Ivkovic, Albert Yeung.
Description: Chicago ; London : The University of Chicago
    Press, 2016. | Includes bibliographical references and index.
Identifiers: LCCN 2016014738 |
    ISBN 9780226338699 (cloth : alk. paper) |
    ISBN 9780226338729 (e-book)
Subjects: LCSH: Stress (Psychology) | Stress (Physiology)
Classification: LCC BF575.S75 F753 2016 | DDC 1555.9/042—
    dc23 LC record available at http://lccn.loc.gov/2016014738

This book was conceived, designed, and produced by
Ivy Press
Ovest House, 58 West Street
Brighton BN1 2RA
United Kingdom
www.quartoknows.com

Publisher Susan Kelly
Creative Director Michael Whitehead
Editorial Director Tom Kitch
Commissioning Editor Kate Shanahan
Senior Project Editor Caroline Earle
Editor Kate Duffy
Designer JC Lanaway
Illustrator Louis Mackay

Gregory L. Fricchione, MD, is one of the world's leading
stress experts. On the faculty at Harvard Medical
School and Professor of Psychiatry, he is Director
of the Division of Psychiatry and Medicine at
Massachusetts General Hospital. He is the author
of over 140 journal articles and of several books,
including *Compassion and Healing in Medicine and
Society* (2011). In 2006 he became Director of the
Benson-Henry Institute for Mind Body Medicine,
one of the top five research institutes for
stress-related illness in the United States.

Ana Ivkovic, MD, is a psychiatrist based at
Massachusetts General Hospital with a specialist
knowledge in nutrition. She is a board-certified
neurologist who trained at the University of Illinois.

Albert S. Yeung, MD, ScD, completed residency training
in psychiatry at the Massachusetts General Hospital
and his major research interests include integrating
primary care and mental health services to improve
treatment of depression, mental health issues
of under-served populations, and the use of
complementary and alternative methods in treating
mood and anxiety disorders. He has authored or
co-authored more than fifty original articles and
book chapters and a book on self-management
of depression.

CONTENTS

INTRODUCTION
How dangerous is stress?

The biggest health challenge facing the world in the twenty-first century is from the effects of stress on individuals and communities. With the exceptions of acute epidemics of infectious communicable diseases such as Ebola, SARS, and influenza, we now recognize that it is the stress-related, chronic non-communicable diseases (NCDs) that are the greatest danger to our mortality, our overall health, and our economy. These diseases include the cardiovascular diseases, chronic pulmonary diseases, diabetes, arthritic diseases, and the neuropsychiatric diseases.

If we are to address this challenge, and hold at bay its consequences, we all must play our part. This is not the sort of health challenge a surgeon can excise or for which a doctor can prescribe a course of antibiotics. All of us—those who are healthy, those who are at risk of future illnesses, and those who are already stricken; as well as doctors, nurses, health psychologists, social workers, and pastoral counselors must pull together in a common effort to reduce the stress-induced drivers that persistently undermine our health. Stress reduces our metabolic reserve and increases the risk of what is called the "metabolic syndrome." The metabolic syndrome consists of obesity, high cholesterol, high blood pressure, and diabetes, as well as chronic, stress-related immunoactivation, which set the stage for the chronic non-communicable diseases.

So, we believe strongly that our communal efforts must begin with public education, and it is in this spirit that we offer this particular book, which sets out to provide the reader with a review of the latest research in the field of stress physiology, along with its disease producing potential. We also introduce a discussion of the antidote—namely the enhancement of resilience. Resilience can be thought of as good adjustment across different domains, which tends to preserve health in the face of significant adversity.

In Chapter One we introduce the topic of stress, explore the history and concept of stress, and examine the effects of good and bad stress on our minds and bodies. Chapter Two takes the exploration of stress on the brain a step further, with an in-depth exploration of how external and internal environmental stressors are processed in the human brain. We look at how the brain senses the world around it and within it, and how it analyzes that experience. We will examine how the brain's chemical messengers respond and are activated by stress-related actions. We also explore the relationship between stress and emotion, cognition, and memory.

Perhaps the strongest evidence for the linkage of mind, brain, and body exists with regard to the brain–heart connection, and this relationship, which is so important for the health of people the world over, is dissected in Chapter Three. We know that coronary artery disease is the number one disease in terms of mortality plus disability, and that depression—a chronic stress disorder—is number two. And, furthermore, these two stress-related diseases feed off one another, with each one raising and increasing the risk for the other.

In Chapter Four, we explore how stress effects a change in the immune system. The immune system is a defense mechanism for our bodies protecting us from external infectious diseases. When under attack from stress, it can break down and become susceptible to stress-related non-communicable diseases.

The often-neglected realm of sleep hygiene is discussed in Chapter Five. It is becoming increasingly clear that stress impedes the capacity to enjoy restorative sleep and without this, one's daily stress response will have a tendency to skyrocket.

In Chapters Six and Seven, we take the reader into two specialized topics in stress research—woman's health and nutrition. The intricate nature of stress physiology and the female hormonal cycle are important factors in our consideration of the stress response in women. And the more we learn about the brain–gut axis and the importance of the microbiota, the more the relationship between our foods and our stressors will need our attention and, thus, we introduce this area of inquiry in Chapter Seven.

In Chapter Eight, the key importance of the social environment in the experience of stress is examined. In the not-so-distant past, scientists wrote books that debated the relative importance for health of nature versus nurture, and heredity versus environment. The recent revolution in our understanding of genetics has minimized this false dichotomy. It has shown that mental resilience can be not only genetic, but can also be positively or negatively influenced by

our environment. The social milieu can be supportive and reduce our fears, but it can also be responsible for the heightened stress response and a variety of traumas that result in post-traumatic stress. This area is examined and reviewed in Chapter Nine. Learning to remain effective in life despite fear is a component of human resilience.

In Chapter Ten, we explore the topic of resilience and its capacity to buffer against the ravages of the chronic stress response. This area of medicine is now commonly referred to as mind–body medicine and integrative health. Whether we live in high-, middle-, or low-income countries, we can all benefit from what this field of medicine can teach us about our health, and how best to optimize it. In this final chapter, we derive a very simple integrative health equation all of us can keep in mind when we desire to stay as healthy as we can in today's stress-filled world. This epidemic of non-communicable diseases accounted for more than 36 million deaths (60 percent) worldwide in 2005, and threatens to create a cumulative output loss of 47 trillion dollars, roughly 75 percent of the global GDP by 2030.

Indeed, most of us today are at risk of developing one or more of the stress-related non-communicable diseases mentioned above. In this regard, educating ourselves, personally and as a society, about stress and what we can do about it through sound self-care approaches, is the first step in taking the action needed to meet the biggest health challenge of the twenty-first century.

GREGORY L. FRICCHIONE, MD

Chapter One

INTRODUCING STRESS

Some of us enjoy the buzz of skiing down a steep mountain, while others thrill to sitting on the edge of our seats nervously watching a horror film. But would it be healthy to experience the excitement of extreme skiing or the spine tingling energy of a movie thriller all day long and every day? The molecules responsible for this extended buzz would soon take their toll on our health.

This chapter introduces the concept of stress and how experiences get processed in the brain to create a sense of tension and demand in the face of challenge and threat. We will explore the conduits that carry the message of stress from specialized brain regions down to our organs and tissues, effectively unifying our constituent parts into a whole organism capable of responding in our most important efforts to survive and prosper. But, if this system gets stuck in a vortex of continual stress responsiveness, a host of chronic stress-related diseases may emerge that can rob us of vitality through the debilitation of illness, and sometimes lead to premature death. Therefore, it is important for us to learn as much as we can about stress and its mechanisms.

LIFE'S JOB DESCRIPTION
What is stress?

The word "stress" carries negative connotations, and undue stress or prolonged stress can have very unfavorable consequences. However, to understand why stress can have such damaging effects, it is important to understand its true, more general meaning. Stress is a living thing's response to changing circumstances in its immediate environment. It is a "sense-analyze-decide-respond" system that is necessary for survival.

All living things have innate mechanisms for sensing and analyzing "stressors." A stressor is anything that could threaten an individual's well-being. In the case of a single-celled organism, a stressor might be a toxic chemical compound. Cells have receptors on their membranes that feed information about the stressors into the cell—and the internal machinery of the cell crudely analyzes the information and the cell responds. For example, a bacterium will move away from toxic chemicals.

Human cells have protective mechanisms at the cellular level, just as bacteria do. But as complex multicellular organisms, we can analyze and respond to complex situations about which bacteria would be unaware. For example, we can read aggressors'

behavior to predict an impending attack and try to move out of the way. For these kinds of stressors, our sense-analyze-decide-respond system is centered on the nervous system. Nerve endings in the skin, eyes, ears, mouth, and nasal passages do the sensing, and convey information to the brain. In the brain, various structures do the hard work of analyzing, deciding, and responding to stressors. The response is carried out via neurons that connect to organs and muscles, or to hormone-producing glands around the body. Generally, a stressful situation will cause the brain to become more alert and prepare the body for action.

SENSE, ANALYZE, AND RESPOND TO STRESS
The brain is at the center of our reaction to stressful events. It receives sensory inputs from the outside world, makes sense of them, and then reacts accordingly: by moving muscles, regulating body systems, and by producing hormones. Most sense inputs feed into a region of the brain called the thalamus. It communicates with the cortex—the wrinkled outer part of the brain—and also with the limbic system, which controls the brain's emotions and drives. The cortex initiates muscular movements, while the limbic system regulates the release of hormones and controls functions such as heart rate and pupil dilation.

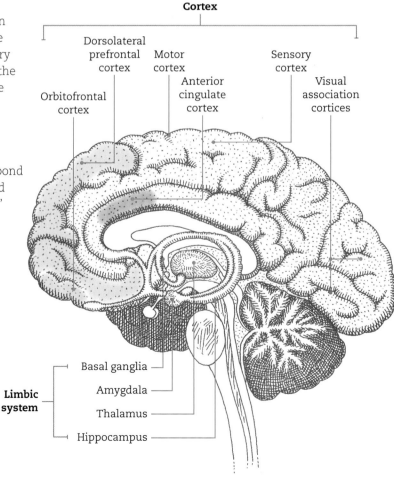

Cortex

Dorsolateral prefrontal cortex

Motor cortex

Anterior cingulate cortex

Sensory cortex

Visual association cortices

Orbitofrontal cortex

Limbic system

Basal ganglia

Amygdala

Thalamus

Hippocampus

Good stress

A little stress is a good thing. Not only can it keep us alive, it can also improve our performance when we carry out a task, or improve our learning of a new skill. This idea was first analyzed scientifically by American psychologists Robert Yerkes (1876–1956) and John Dodson (1879–1955). They measured the increases in performance in people who had been presented with challenges, and whose brains and bodies had become alert and ready for action, or "aroused." Their findings can be summarized as a simple curve of arousal versus performance (see below). The state in which a living thing is functioning well and dealing with—or even benefitting from—a variety of everyday stressors is called "eustress."

Bad stress

However, not all stressful situations are good—and not all stress is beneficial. When there is too much arousal, or too much pressure of one kind or another, performance suffers.

When the challenges of life escalate and we feel threatened, we are often faced with what is called toxic stress or "distress"—the opposite of eustress. When we do not have enough to eat, lose our jobs or homes, or our relationships falter, we experience emotions and have thoughts that are accompanied by physiological and hormonal markers of our disquiet. In addition, we can fall victim to internal drivers of our distress. For example, we may worry about future challenges. Feeling anxious about whether or not we can meet these challenges, and brooding on past failures, can further exhaust us with toxic stress. So, for many reasons, our system may come up short, and this is when distress can set in.

THE YERKES-DODSON LAW

The Yerkes-Dodson Law (1908) seems to apply in sports, academic pursuits, the arts, and even social situations. It relates arousal (stress) to performance. Here, arousal is a state in which a flood of hormones increases muscle tone and heart rate, and heightens senses. It is achieved whenever the body is physiologically or mentally challenged, or stressed. The relationship between arousal and performance is best understood by looking at the chart below. Optimal performance is achieved with a moderate amount of arousal, while performance is less than optimal if there is little or no arousal—and if there is too much. As we shall see in this book, not only can a high level of stress reduce performance, it can also lead to physical distress and illness.

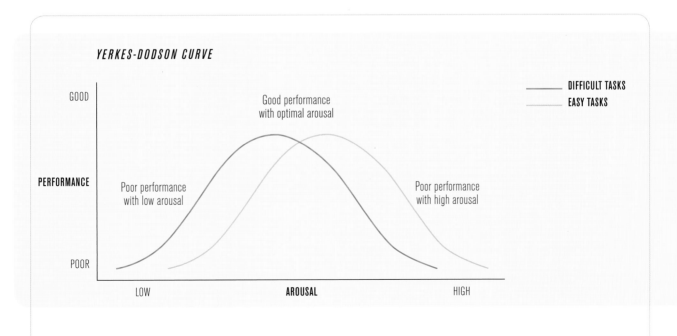

YERKES-DODSON CURVE

——— DIFFICULT TASKS
——— EASY TASKS

GOOD

Good performance with optimal arousal

PERFORMANCE

Poor performance with low arousal

Poor performance with high arousal

POOR

LOW   AROUSAL   HIGH

THE HISTORY OF STRESS
What does stress mean?

Dictionary definitions of stress define it as:

"a state of mental or emotional strain or tension resulting from adverse or very demanding circumstances"

or

"something that causes such a state."

Stress is also described as

"a physical, chemical, or emotional factor…to which an individual fails to make a satisfactory adaptation, and which causes physiological tensions that may be a contributing cause of disease."

Elsewhere stress has been defined as

"a threat, real or implied, to the psychological or physical integrity of an individual."

Homeostasis: Maintaining internal stability
The first description of stress as a concept is commonly attributed to the French physiologist Claude Bernard (1813–78). In 1865, he presented the idea that an organism is comprised of an internal environment made up of cells called the *milieu intérieur* or "interior milieu." This milieu is tightly controlled through a series of feedback mechanisms based on information flowing in from the external environment. Then, in the early twentieth century, American physiologist Walter B. Cannon (1871–1945) introduced the concept of "homeostasis," by which he meant the way in which the body maintains its internal equilibrium even when it is faced with external difficulties.

In his study of one major part of the response to stress, namely, the sympathetic nervous system (SNS), Cannon came to realize that organisms adapt by compensatory responses when faced with challenges to their internal stability. For example, missing your bus with the threat of being late for an important business meeting will lead to a whole host of internal physical changes, many brought about by the sympathetic nervous system. This system has its source in one of your brain stem nuclei called the locus coeruleus—the blue location—named because its melanin granules stain blue in the laboratory. From there, sympathetic nerve fibers track to your lateral hypothalamus and down the spinal column to exit into nerve bundles called ganglia, which then supply your organs and muscles with action responses. An acute challenge to your well-being will ignite an outpouring of energy-demanding chemical messengers (see also page 42) which enable you to make every effort to survive in, what Cannon dubbed, a "fight-or-flight reaction." But when the challenge stops, your body is expected to restore a homeostasis (return to equilibrium) with a different chemical profile that is more in keeping with a sense of security, and energy conservation.

The term "homeostasis" is meant to emphasize that organisms can maintain a multitude of physiological variables such as blood glucose, oxygen tension, blood pressure, heart rate, and core temperature, within acceptable ranges for health. This feature of successful living requires the presence of feedback systems. Sensors are necessary to gauge when physiological values are out of bounds, and effectors are necessary to bring them back to normal. Think of your thermostat at home. It is a homeostatic device in a way. When it senses a temperature plunge, a feedback loop ignites your heating system to maintain a temperature homeostasis in your home.

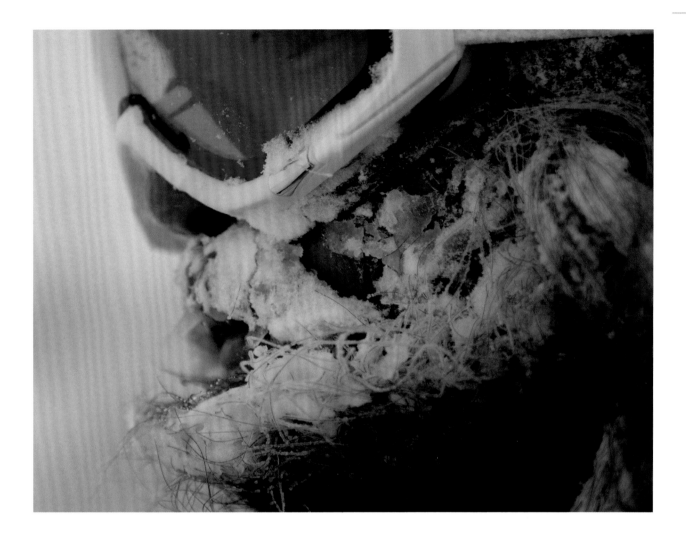

In human beings, when we get cold, shivering commences and blood is redirected from surface vessels back to inside the body. On the other hand, when our core temperature rises, we sweat and blood is shunted from our internal organs (viscera) to the skin so that heat can escape. It is felt that many threats to homeostasis, such as intense cold, blood loss, low glucose, trauma, and psychosocial distress, would all induce a response from the fight-or-flight system, which comprises the sympathetic nervous system and the medulla part of the adrenal gland, in an effort to restore our safety. This so-called adrenergic system runs on neurotransmitters called catecholamines (epinephrine and norepinephrine, also called adrenaline and noradrenaline).

COMING IN FROM THE COLD
The stress of the cold temperature has affected the mountaineer's physiological response. He has maintained homeostasis by shunting blood flow to his internal organs, while at the same time reducing skin capillary blood volume on the surface, and conserving energy by reducing as best he can highly charged emotional displays.

Neurotransmitters are chemical messengers that exchange information through actions at post-synaptic receptors that establish (binding) connections between neurons. Whether we change posture, eat a dinner with gusto, tackle a mugger, or need to give a speech in front of a group of strangers, our nervous system will be activated through the actions of these neurotransmitters.

THE CONCEPT OF STRESS
What is the HPA axis?

Hans Selye (1907–82) was a Hungarian scientist, who in 1956 popularized the concept of stress. The word "stress" originates from the Latin word *strigere* meaning to tighten. Selye adapted the term, which was used in engineering to denote a deforming force that results in structural strain, and he applied it to the organism's response to internal and external disturbances called "stressors." Stress became "the nonspecific response of the body to any demand upon it."

No matter what type of stressor an organism encounters, Selye contended that the body's stress response is liable to be composed of a set of similar features. This became known as "the non-specificity hypothesis." Selye's concept stood in opposition to a psychoanalytic concept that specific stressors gave rise to specific stress responses culminating in specific diseases.

THE FIGHT-OR-FLIGHT RESPONSE

The trauma of being in a war zone is one of the most extreme stressors that a person can experience and many combatants go on to develop post-traumatic stress disorder (PTSD) when their acute, war-induced stress response persists past three months. The stereotypical example of fight-or-flight stress is depicted in this classic painting, *The Death of Major Peirson, 6 January 1781* by John Singleton Copley.

The HPA axis

Selye proposed three phases involved in coping with stress: first, an alarm reaction strikes us, not unlike the concept of the fight-or-flight response; second, we adapt to the stressor, often by resisting it and seeking a return to homeostasis; in the third stage, we run the risk of exhaustion if the stressor is prolonged or catastrophic. Our resistance to stress owes to a major stress response system comprised of three hormonal nodes: the hypothalamus (H), the pituitary (P), and the adrenal cortex (A), called the "HPA axis" for short. The output glucocorticoid hormone is called cortisol in the human. This extremely important stress hormone contributes to stress resistance and a return to homeostasis, but can also provoke illness when its production is excessive or persistent.

Selye proposed that the body could change its set point to a new steady state goal in an effort to resist unusually high demands. He called this process "heterostasis" (*hetero* from the Greek for "other"); and he felt that mind–body stress reducing approaches, which enhance our body's natural defenses, while not curative, can have the benefit of changing the set point and maintaining health. Today, we can think of integrative mind–body approaches as heterostatic in their effects.

HOW THE HPA AXIS WORKS

When the amygdala perceives sensory information from the thalamus to be threatening, it engages the paraventricular nucleus in the hypothalamus resulting in the stimulation of corticotropin-releasing hormone (CRH), which begins the stress hormone cascade. This hormone then stimulates the pituitary to release another hormone called adrenocorticotrophic hormone (ACTH). This hormone travels down to the adrenal cortex gland, which produces the stress hormone cortisol. Cortisol in turn will feedback to the hypothalamus and the pituitary.

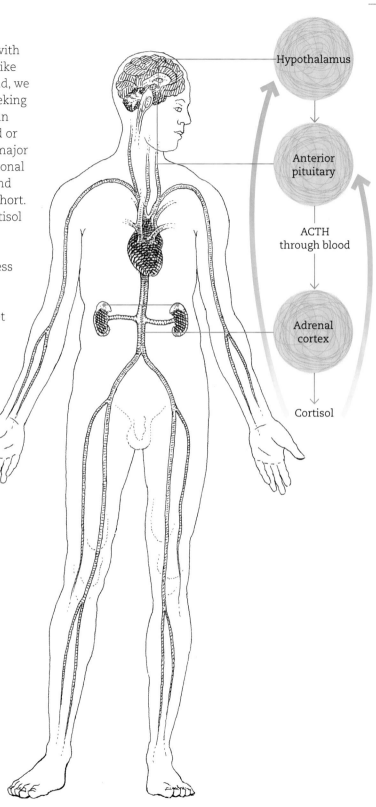

Hypothalamus

Anterior pituitary

ACTH through blood

Adrenal cortex

Cortisol

THE BIOLOGY OF COPING WITH STRESS

What happens to the body under stress

In the latter part of the twentieth century adjustments needed to be made to Selye's concepts of stress, as understanding had developed. First, it became clear that the idea of a uniform general stress response, regardless of the specific nature of the stressor, needed refinement. A stressor—for example extreme cold—will stimulate your sympathetic nervous system while allowing your other hormonal responses to remain relatively dormant.

It has also been recognized that homeostatic physiological systems are multiple, interactive, and dynamic, responding in a real-time way to external and internal fluctuating demands. Stress reflects a discrepancy between expectations, which are products of your genetic predispositions, developmental learning, and analysis of the present state of affairs, and your anticipated perception of environmental demands. This discrepancy leads to patterned and compensatory responses on your part.

Today, many researchers conceive of stress as the perceptual aftermath of a threat to your security and homeostasis. The stress response to specific stressors has both specific and non-specific components. Variables include the nature of the challenge to your homeostasis, your appraisal of the stressor, and the optimism or pessimism you associate with the task of coping with it.

Allostasis

Physiologists (Sterling and Eyer in 1988) coined the term "allostasis" to capture this new notion of stress-responding physiological systems in the case of blood pressure control. Allostasis, as researcher Bruce McEwen points out, means "maintaining stability, or homeostasis, through change." Thus, changing environmental conditions leads to variable physiological and hormonal states that enable you to meet your separation challenges in a safe and effective way without wild or persistent fluctuation from the set point. A good example of allostasis is the well-known finding of heart rate variability. The heart's physiology is considered healthier when the beat-to-beat heart rate interval, as measured in the electrocardiogram, fluctuates over time in response to environmental demands. A heart that does not alter its contraction rate reveals pathology in its lack of dynamic stability, adverse cellular dynamics, and poor evaporative mechanisms.

Moderate exercise permits allostatic mechanisms to predict your metabolic needs and to meet them in a balanced fashion. Excessive exercise, on the other hand, can lead you to experience a host of negative physiological changes by producing an allostatic challenge. Your body might then enter a catabolic breakdown state rather than an anabolic build-up state. You may become dehydrated, low in glucose and oxygen saturation. Cells may begin to undergo anaerobic (without oxygen) respiration, which releases lactate and leads to a phenomenon known as lactic acidosis. This can cause spasms and aches in your muscles, lack of concentration, general fatigue and shortness of breath. In this context, muscle cell can breakdown, which can lead to kidney dysfunction. Proteins and calories may rapidly deplete leading to deficiencies. Cortisol may persistently elevate, resulting in vulnerability to mood changes and infections. Heart rate variability may decline and the body may begin to feel tired as fatigue sets in.

Three typical responses to stress

The allostatic process is supervised by our brain and requires a large expenditure of energy. When someone is threatened by overwhelming stress, for example from experiencing an acute traumatic event such as a bomb exploding, or a more persistent stressor like losing your home, tremendous metabolic strain will lead to disease vulnerability. We suffer "metabolic wear and tear" if we are forced to continually expend energy when responding to repetitive stress. "Allostatic loading," as it is also known, can be thought of as the price the body pays for being forced to adapt to adverse psychosocial or physical situations.

As opposed to the idea that the stress response will always be the same, the concept of allostasis allows for an element of specificity in your stress response. For example, the predominant specific responder system to activate in situations of blood pressure changes will be the sympathetic nervous system. In contrast, when you experience a deprivation of glucose or emotional distress, the adrenal medulla will be the predominant responder.

However, three general allostatic stress responses can be recognized. First, stressor and response may be well-matched resulting in a return to homeostasis (resilience through allostasis). Second, an overblown, continual response may cause us to veer off into danger zones (vulnerability due to allostatic loading). And third, the match may be optimal leading to a new, stronger set point (post-traumatic growth or anti-fragility). Post-traumatic anti-fragility emerges when, though the threat is enormous, we find within ourselves the strength to emerge in a wiser and stronger position.

NETWORK SUPPORT

Facing the constant threat of wartime attack taxes your allostasis monitoring system. Your protective sensory-motor analyzer-effector system is constantly on-call using up valuable energy to power your amygdala's hypervigilance and to maximize your cortex's strategies to stay calm and effective in the midst of fear. This metabolic wear and tear comes with the brain's efforts to maintain allostasis and leads to a host of illness vulnerabilities. Returning home to support, attachment, and love will reduce stress and promote overall health and a happier future.

BIOLOGICAL ATTACHMENTS
The need for nurture

Amphibians and reptiles are born raring to go and require little parenting to get a good start in life. We humans are born in a helpless state and rely on nurturance for our early survival. This stark difference in the early developmental strategies has a far-reaching impact on what constitutes stress in our mammalian species. Indeed, there is no species on earth that is born more helpless (altricial) than we are. It is not surprising then that attachment insecurity may trigger our disease pathways by increasing our stress.

Survival strategy

Our survival strategy is therefore locked in and plays itself out in the evolution of our brain structures and functions. Simply stated, we have a much better chance of survival when we are nested in the safety of caring parents and social group members. At a very basic level our brains will mature in a much healthier way when our upbringing matches our evolutionarily based needs with regard to basic attachment to our parents and caregivers. As the English psychiatrist and father of Attachment Theory, John Bowlby (1907–90), said: "Our environment of evolutionary adaptedness is one of secure base attachment."

Stress, which previously in evolutionary history related to threats to individuals and difficulty in obtaining food and sexual partners, now also became associated with threats of separation from parents or children, and from social supports. Therefore, it is not surprising that attachment insecurity may trigger our disease pathways by increasing our stress and engaging in an inflammatory response. This activation of the immune system is a remnant of the first survival challenge shared by all vertebrates—namely microbial infection. It is appropriated later when social stress becomes an issue as a defensive strategy that comes with a cost.

Frontal brain regions

Our frontal brain regions, which receive stress-laden signals from the amygdala, can reduce the outpouring of energy-depleting glutamate surges by using the brain's most important inhibitory neurotransmitter—gamma-aminobutyric acid (GABA). Important among these frontal regions is the anterior cingulate cortex. It is here that the message of separation pain is especially felt and the decision to act to achieve an attachment solution is crafted, taking into account a bounty of cognitive and emotional information. When you perceive yourself as once again securely attached, the anterior cingulate and other medial prefrontal cortical regions will reassure your amygdala that the threat to your attachments has been thwarted.

We all face the separation challenge, and it may form the foundation for all future anxieties and fears. It was there first when we evolved our survival strategies as mammals, and it is there first when we develop individually as infant human beings. Thus, it is important to view stress and the stressors that precipitate stress using attachment theory.

Attachment anxiety

Attachment anxiety is associated with distress and leads to unhealthy reductions in heart rate variability, and increases in blood pressure, and stress hormones, and other illness vulnerabilities. Insecure childhood attachments, especially those brought on by trauma, may thus predispose to illnesses later in life. In addition, insecure attachment style may lead to the use of maladaptive behaviors, such as smoking and drinking, which are illness-provoking. Conversely, the bolstering of attachment security through social support and compassionate love may be healthful. This experience buffers against the metabolic wear-and-tear effects of our stress and elevates our disease threshold, making it less likely we will fall ill.

RESPONDING TO ANXIETY

Imagine that you are a parent whose three-year-old child wanders off during a shopping trip to the mall. You are frantic with worry. The amygdala, which is responsible for the generation of fear, ignites with distress sending pulses of the brain's most plentiful excitatory neurotransmitter, glutamate, surging throughout the brain. Destinations include the locus coeruleus and the hypothalamus resulting in activation of the sympathetic nervous system. Another destination is a portion of the hypothalamus called the paraventricular nucleus, which results in the HPA axis stress response (see page 15). Finally, a third part of the hypothalamus called the mediobasal nucleus sets in motion an inflammatory response in the brain, even without an infection to set it off. Thus, the stress of separation alone can make you feel like you have an infection or illness.

Of course not only is the separated parent prone to enormous stress. Even securely attached children, when they realize that they are separated from a parent, will experience their amygdala sounding the alarm. This response will have a tendency to be at its most frightening for children who, in the first place, face separation from a position of insecure attachment.

SEPARATION AND STRESS

Infants in the first 18 months tend to feel psychologically united to their mothers. In a process called separation-individuation, children from 18–24 months practice spending longer periods at greater distances from the mother before running back. Soon children recognize that their mother exists for them even when not seen or heard. This reduces separation stress and advances children's individual development. But, this original separation anxiety is always with us. This is why the image of a lost child conjures up in us a bundle of harrowing thoughts and feelings.

THE STRESS RESPONSE SYSTEM
What physically happens when I feel stressed?

We may find ourselves in the wrong part of a city late at night facing a menacing gang, or we may need to make a big presentation in front of our bosses, or we may anticipate an argument over finances with our spouse. Our brains appraise the situation with our senses. What was my last experience of this? How did it turn out when I did A, B, or C? Can I cope now? When there is a question mark in our minds and we perceive a threat to our safety, we may enter into stress mode.

A myriad of actions ensues. First, our bodily energy shifts to acute needs from long-term ones such as tissue growth, digestive processes, and sexual functioning. Certain action-oriented muscles require more energy and lungs need more oxygen. Pain sensitivity is dulled and our tendency to bleed is curtailed. In order to set these prioritized actions in motion, the fear-conditioned amygdala engages the hypothalamus.

The stress response occurs along the HPA axis. This means the amygdala-stimulated hypothalamus excites the pituitary, which alerts the adrenal glands atop the kidneys. The adrenal medulla responds by pouring out a catecholamine called epinephrine (adrenaline). This transmitter sets the fight-or-flight response in motion. Pulse rate bumps up sending more blood glucose and oxygen to muscles and lungs, and also to the brain itself, to keep us more alert. Blood vessels constrict and fibrinogen is produced to help with clotting. Epinephrine facilitates the conversion of glycogen to glucose and breaks down and releases fatty acids from fat stores to supply us with a ready source of energy.

A second wave of defense then emerges through the HPA axis, causing the outpouring of the stress hormone cortisol. Cortisol's initial role is to build back up the energy reserve depleted by the epinephrine surge. It does this by converting foodstuffs into energy storage in the form of glycogen and fat.

AUTONOMIC NERVOUS SYSTEM

The stress response also engages the autonomic nervous system. It prepares you to surge into action to meet an environmental challenge to your attachments. In this response, your heart speeds up, blood pressure rises, pupils dilate, and digestion slows. Finally, the parasympathetic nervous system, mainly through the auspices of the great vagus nerve, dampens your autonomic nervous system and restores the nervous system to its normal state, hopefully before too much energy is wasted. If used too frequently this system can contribute to a variety of stress-related diseases, especially cardiac disease.

A TWO-STAGE RESPONSE
After the initial epinephrine (adrenaline) rush along the HPA axis, further hormones such as corticotropin (CRH) are released in order for your body to stabilize function in the face of change.

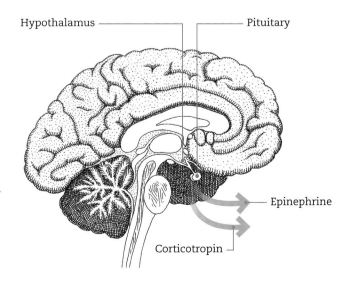

Hypothalamus — Pituitary

Epinephrine

Corticotropin

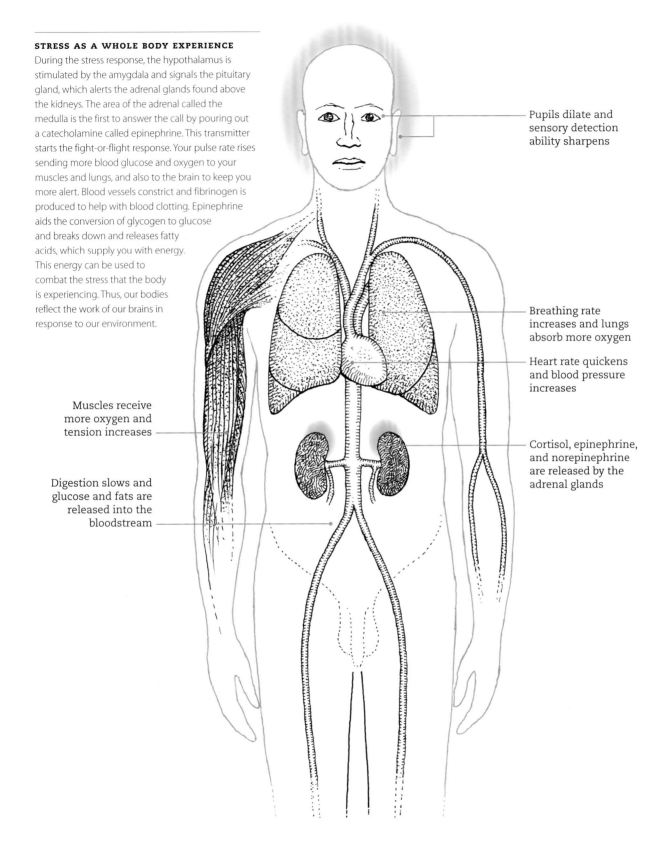

STRESS AS A WHOLE BODY EXPERIENCE

During the stress response, the hypothalamus is stimulated by the amygdala and signals the pituitary gland, which alerts the adrenal glands found above the kidneys. The area of the adrenal called the medulla is the first to answer the call by pouring out a catecholamine called epinephrine. This transmitter starts the fight-or-flight response. Your pulse rate rises sending more blood glucose and oxygen to your muscles and lungs, and also to the brain to keep you more alert. Blood vessels constrict and fibrinogen is produced to help with blood clotting. Epinephrine aids the conversion of glycogen to glucose and breaks down and releases fatty acids, which supply you with energy. This energy can be used to combat the stress that the body is experiencing. Thus, our bodies reflect the work of our brains in response to our environment.

Pupils dilate and sensory detection ability sharpens

Breathing rate increases and lungs absorb more oxygen

Heart rate quickens and blood pressure increases

Cortisol, epinephrine, and norepinephrine are released by the adrenal glands

Muscles receive more oxygen and tension increases

Digestion slows and glucose and fats are released into the bloodstream

DISTRESS
When stress becomes chronic

Stress as *distress* has been defined as a state of disharmony or of threatened homeostasis. It refers to a disruption of the dynamic equilibrium among a person's physiological, psychological, and social dimensions, because of the perceived presence of an external or internal threat. Alterations in the external or internal environment may provoke a physiological response mediated by several interconnected physiological systems constituting the stress response.

The acute stress response costs us energy and so it is crucial that our brain and body does not remain in chronic threat mode. When it does, too much cortisol can harm our body. It can block insulin receptor action and make it harder for muscle to absorb glucose. It can enhance the storage of energy as fat in the abdomen, which then poses a health threat, partly because this fat can induce inflammation (see opposite). In addition, it can convert muscle protein to fat. Cortisol can also cause our bones to demineralize and lose calcium setting the stage for later osteoporosis as well.

Cortisol affects our immune system. When there is a chronic overabundance of cortisol, immune function can be suppressed making us more susceptible to infections. Acute cortisol surges, on the other hand, can help activate our white blood cells, direct them to sites of acute injury and, once there, help them to stick to the walls and tissues where they can help fight infection and promote wound healing. The brain also uses cortisol to signal when an acute immune response, sometimes called "an acute phase response," should come to a close. Sometimes receptors for cortisol stop functioning from overuse and the stress response can become chronic, setting the stage for a variety of autoimmune conditions, for example lupus and lupus-like syndromes, multiple sclerosis, and rheumatoid arthritis.

When the stress hormones, epinephrine and cortisol, are poorly regulated by our brain due to overwhelming or repetitive stressors, our body pays a price in terms of allostatic load. So, too much epinephrine predisposes to hypertension and sets the stage for atherosclerotic disease. And too much cortisol predisposes to obesity and type 2 diabetes.

The brain may also become vulnerable to dysfunctional stress hormone states. This may be due to overproduction or underproduction and also to abnormal timing of the normal daily circadian rhythms of epinephrine and cortisol flows in the body. For example, abnormal patterns of cortisol secretion occur in depression and in post-traumatic stress. In addition, the dysfunctional secretion of cortisol may conspire with abnormal functional activity in stress-mediating parts of your brain, such as the amygdala and the hippocampus, and lead to brain cell atrophy and death. These brain regions are crucial for learning, memory, and the immune response.

The stress response will also engage our autonomic nervous system. It is activated to meet external stress demands through the action of the sympathetic nervous system, which helps us to deal with environmental challenges. The autonomic nervous system also provides for a return to the way things were before the threat had passed through the parasympathetic nervous system. A chronic stress response will also lead to persistent elevation of catecholamines (epinephrine and norepinephrine) and this will make us more vulnerable to hypertension with its related illnesses.

Cortisol helps to maintain homeostasis in the face of stress. It helps to energize us, and regulates our blood pressure and insulin release. It can improve our energy, cognition, and immune response while diminishing our pain sensation.

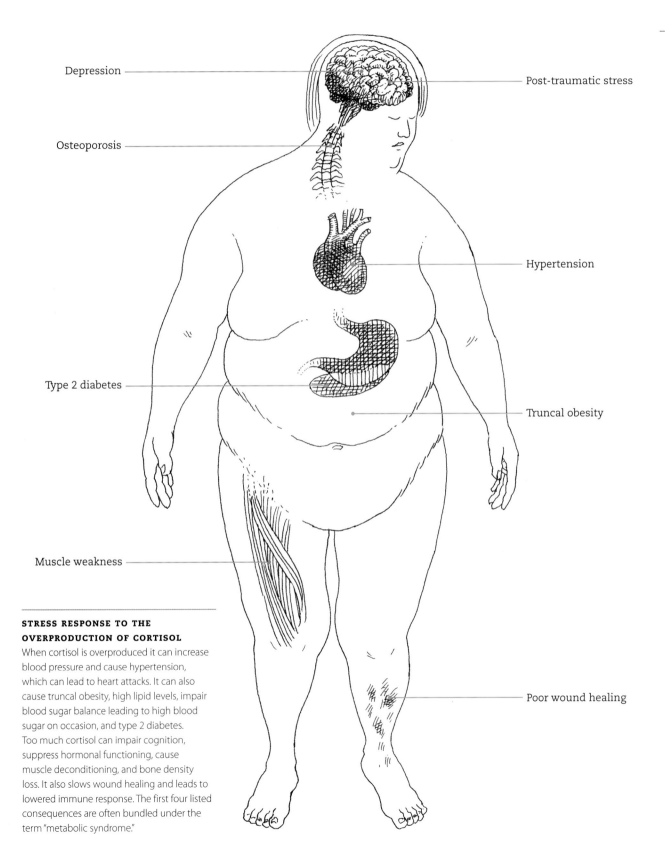

Depression

Osteoporosis

Type 2 diabetes

Muscle weakness

Post-traumatic stress

Hypertension

Truncal obesity

Poor wound healing

STRESS RESPONSE TO THE OVERPRODUCTION OF CORTISOL
When cortisol is overproduced it can increase blood pressure and cause hypertension, which can lead to heart attacks. It can also cause truncal obesity, high lipid levels, impair blood sugar balance leading to high blood sugar on occasion, and type 2 diabetes. Too much cortisol can impair cognition, suppress hormonal functioning, cause muscle deconditioning, and bone density loss. It also slows wound healing and leads to lowered immune response. The first four listed consequences are often bundled under the term "metabolic syndrome."

PHYSIOLOGICAL STABILITY AND CELLULAR STRESS

How stress affects the cells

It is now increasingly clear that stress can get translated into metabolic activation and processed at the cellular level as *oxidative stress*—the cellular strain that comes from metabolizing an overabundance of oxygen. Mental stress reflects a challenge to our allostatic state of "stability through change." It takes metabolic energy for our brains to maintain physiological parameters within a normal range in response to external and internal environmental stressors. If our energy expenditure is too high, this can lead to allostatic loading. This stress hormone- and transmitter-induced state alerts target tissues to the need to alter their metabolisms to meet this new state of affairs in order to maintain allostasis. This works well if the challenge or threat is acute and self-limited. However, if the psychosocial stress is chronic, this process can make us vulnerable to disease because of the cellular oxidative stress.

Cellular oxidative stress simply put means that the mitochondria in cells are required to metabolize more oxygen when the person is in the throes of elevated or persistent stress. Therefore, in response to persistent stress, our cells have to work harder, producing more by-products that may cause us long-term damage.

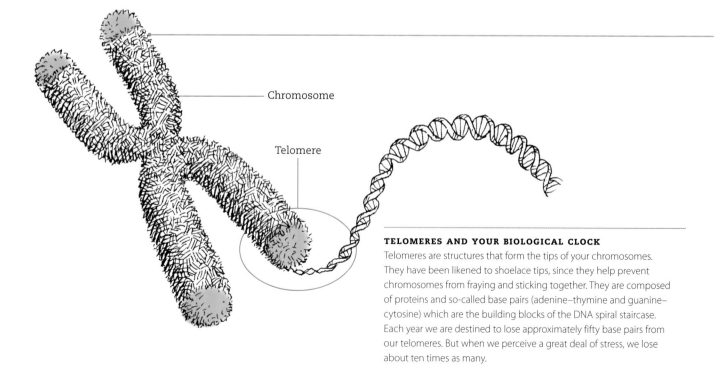

Chromosome

Telomere

TELOMERES AND YOUR BIOLOGICAL CLOCK
Telomeres are structures that form the tips of your chromosomes. They have been likened to shoelace tips, since they help prevent chromosomes from fraying and sticking together. They are composed of proteins and so-called base pairs (adenine–thymine and guanine–cytosine) which are the building blocks of the DNA spiral staircase. Each year we are destined to lose approximately fifty base pairs from our telomeres. But when we perceive a great deal of stress, we lose about ten times as many.

MITOCHONDRIAL FISSION

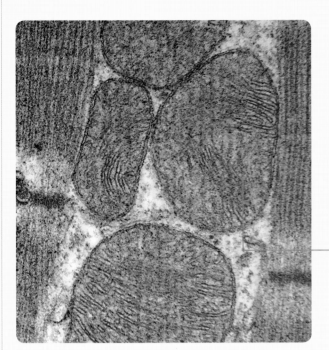

MITOCHONDRIAL SEPARATION STRESS
Mitochondria are tiny organelles inside every cell that release energy for the cell to use. This transmission electron micrograph (TEM) shows normal mitochondria in a muscle cell. Severe stress can cause mitochondria to split apart, in a process known as "mitochondrial fission." In this state, mitochondria produce less energy and are programmed by the cell to die off. Under very favorable, low-stress conditions, mitochondria can join together, or fuse, forming long continuous filaments.

—————— Mitochondria

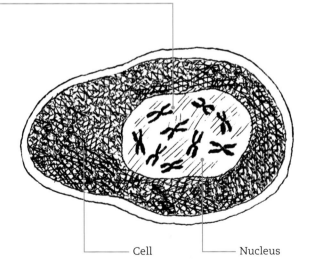

—— Cell   —— Nucleus

Telomeres
In an important study that examined how psychosocial stress can lead to cellular stress, Epel and her colleagues (2004) were able to show that the stress perceived by mothers taking care of their chronically ill children was associated with white blood cell shortening of telomere length. Telomeres are the tips of chromosomes comprised of base pairs of your DNA surrounded by a material called chromatin. Normally, as we get older we lose about 50 base pairs a year from our telomeres. But when our stress is high, as with these mothers, upward of 550 base pairs were lost in a year! Furthermore this finding was associated with increased oxidative stress. As mentioned oxidative stress refers to the fact that your cells have to process more oxygen and glucose under conditions of stress, and this results in the overproduction of metabolites that are harmful to the cell.

THE EPIGENETICS OF STRESS
The effect of nature and nurture

It is becoming increasingly clear that environmental influences can activate and deactivate some of your genes, which can result in illness. This field of study has become known as *epigenetics*. Research in this area is proving that there is a very close connection between heredity and environment, nature and nurture. In an important animal study, it was shown that young rats raised with poorly nurturing mothers had reduced activation of a gene that behaviorally made these poorly nurtured rats grow up to be highly susceptible to stress.

The future of stress research
Persistent elevation of stress mediators (cortisol and the catecholamines) may be responsible for an epidemic the world is facing. This epidemic is called *the metabolic syndrome*. This means that stress-related hypertension, high cholesterol, poor insulin receptor functioning, and truncal obesity contribute to development of coronary heart disease, other atherosclerotic illnesses, and type 2 diabetes. The world is also facing what has been called the twenty-first-century's major health challenge, namely stress-related non-communicable diseases (NCDs). These include cardiac and chronic pulmonary disorders, diabetes, arthritides, cancers, and neuropsychiatric problems. The metabolic syndrome is endemic in both high-income and low-income societies and is the precursor to the major non-communicable diseases. Both the metabolic syndrome and the non-communicable diseases can emerge in the setting of heightened stress responses. This is why we need to understand the stress response in the brain.

HOW STRESS-INDUCED FEAR AFFECTS THE BODY

CENTRAL NERVOUS SYSTEM
- Perception becomes narrowed
- Memory becomes imprecise
- Learning is blocked
- Conditioning is defense
- Tendency to regress or repeat responses
- Expectancies are negative
- Tone is to flee or fight

MUSCULAR SYSTEM
- Tension in muscles
- Ready for action
- Jaws clench
- Body braces for action

AUTONOMIC NERVOUS SYSTEM
- Heart rate increases
- Blood pressure rises
- Oxygen need is greater
- Breathing rate increases
- Palms and face sweat
- Blood sugar imbalance
- Epinephrine flows
- Digestive tract shunts blood to muscles
- Blood vessels constrict in hands and face and clotting is enhanced

MANAGING YOUR STRESS RESPONSE

You can help to decrease your stress response and build up your resilience to stress by using mind–body techniques such as those listed below:

- Mindfulness meditation
- Spiritual connectedness
- Relaxation response breathing exercises
- Mindful exercise such as yoga or tai chi
- Moderate exercise
- Balanced nutrition and healthy eating
- Good sleep hygiene
- Social support and pro-social behavior
- Cognitive skills development to overcome automatic negative thoughts and the negative emotions that follow
- Positive psychology approaches such as optimism, purpose, meaning, and gratitude
- Healthy habits and avoidance of health risk behaviors

MEDITATION CAN ALLEVIATE STRESS

A variety of meditative approaches can reverse the stress associated physiological changes that confer illness vulnerability. They reduce sympathetic nervous system drive and increase parasympathetic tone resulting in lowering of blood pressure, heart rate, and respiratory rate. Oxygen consumption also decreases. Research also now shows the gene expression in your white blood cells changes to a healthier pattern reflective of reduced oxidative stress and reduced chronic inflammatory response. This may be responsible for the general health promoting effects of regular meditation.

Chapter Two

STRESS AND THE BRAIN

When we think or experience something, our brain chemistry changes. Certain genes are activated and others deactivated, leading to the production of proteins that will have structural and functional effects on our neurons. These effects incrementally adjust the way networks of neurons function. Thousands upon thousands of these functional adjustments can amount to changes in our behavior. When our brain is stressed, a signature brain pattern emerges which produces the emotional mind experience we associate with being stressed. In this chapter, after presenting the concept of the brain as a physiological monitor, we examine how stress-related actions are produced, and explore the relationship between stress and emotion, cognition, memory, and aging, before looking at brain messengers and other aspects of the brain under stressful conditions.

THE MIND AND THE BRAIN ARE ONE
The mental and physical responses to stress

When we are occupied by challenges, threats, and stressors of one sort or another, the chemical elements of our brain will change, and this change will affect the brain's stress response systems. This realization leads to an approach to medical care that has been dubbed the "bio-psycho-social approach." In this approach, it is understood that all health and illness result from the interplay of biological, psychological, and social factors, which are intimately woven together into a mind–brain–body unity, which we call the whole person, and contributes to whole-person health and well-being.

Our social stressors can cause discomfort just as our physical injuries can be painful. It is known from imaging studies that our brain mediates these effects in association with activations in the same areas, particularly in the part of the brain called the anterior cingulate cortex.

Thus, biological disease does not occur in a vacuum. When our psychological state is precarious due to stress (for example, a fearful father who is constantly worried about the safety of his children); and if our social situation is unsupportive and additionally stress inducing (for example, a boss bullying a single mother with financial challenges), then on a cellular level, we will become increasingly vulnerable to disease and illness. The cells of our brain and body must expend tremendous energy in an effort to stabilize our physiologies in the face of distressing psychological and social stressors.

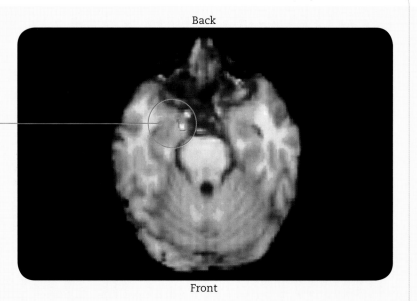

THE BRAIN'S FEAR RESPONSE

Amygdala activity

Back

Front

FEAR FACTOR
A PET (colored positron emission tomography) scan shows a fear response in the brain, see the area marked red and yellow. This image shows a transverse section through a human brain. The active brain region in red and yellow is the left amygdala (at upper center left). The amygdala is "stress central" in our brain.

SOCIAL PAIN

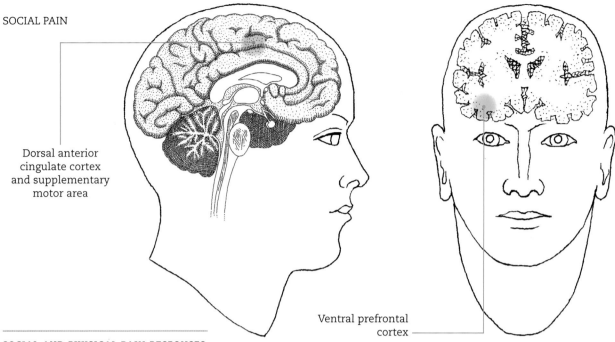

Dorsal anterior
cingulate cortex
and supplementary
motor area

Ventral prefrontal
cortex

SOCIAL AND PHYSICAL PAIN RESPONSES
Social discomfort and physical pain
produce similar responses in the
brain. The anterior cingulate cortex
plays a key role in the physical–social
pain overlap. The right ventral
prefrontal cortex helps to mediate
the distress signal.

Ventral prefrontal
cortex

PHYSICAL PAIN

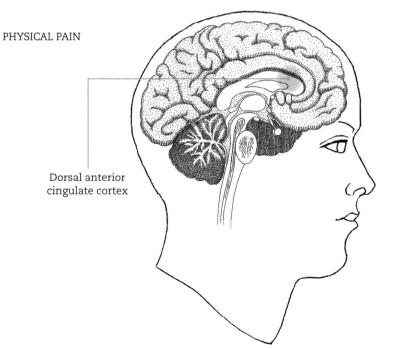

Dorsal anterior
cingulate cortex

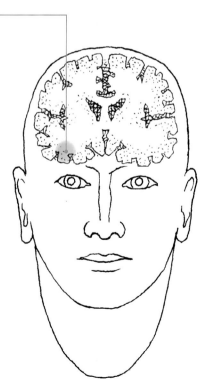

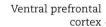

THE BRAIN AS MONITOR OF PHYSIOLOGICAL STABILITY

How the brain responds to physical stress

Our brain actively maintains physiological stability in the face of changing circumstances in a process called allostasis, as we learned in Chapter One. It anticipates increases in energy demand and adjusts our stress response systems to accommodate. When stress is overwhelming or constant, it will inflict metabolic wear and tear (allostatic loading) on our cells. In such a situation, our brain finds it difficult to maintain energy balance. This will place our health in a vulnerable position.

Occasionally, our brain cannot maintain the internal balance for various reasons. As discussed earlier, Selye (1946) described the so-called General Adaptation Syndrome by identifying three stages: the *alarm stage*, in which the acute stress response is activated; the *stage of resistance*, in which the body makes an effort to return to its previous state of stability (homeostasis), but the perception of a threat is still present causing the stress response to persist; and the *state of exhaustion*, in which the stress continues for a long time, the body cannot function normally, and organ systems fail.

Stability through change

Mental stress challenges our physical allostatic equilibrium. It takes added metabolic energy for our brain to maintain normal physiology under conditions of chronic or high stress. It is the brain's stress response systems that serve to notify our body's end organs of challenges or threats perceived by our brain. These end organs (for example, heart) and target tissues (for example, muscle) also alter their metabolism in order to remain stable. This teamwork works well if the challenge or threat is acute and self-limited, but if chronic or overwhelming, this process can result in illness.

THE STATE OF EXHAUSTION

The state of exhaustion is associated with high wear and tear on the body, which stems from the persistent overactivation of our stress response systems. There are four proposed mechanisms associated with this wear and tear (allostatic load):

1. Frequent stress or multiple stressors;

2. Prolonged exposure to stress complicated by a lack of adaptation;

3. Slowness or inability to shut down allostatic responses once a stressor is terminated;

4. An inadequate (insufficient) response, leading to overactivation of other systems in compensation for the deficit.

Oxidative stress

Research shows that psychosocial stress can, thus, get translated into metabolic overactivation at the cellular level, leading to what has been described as "oxidative stress" (for more information see page 24). Oxidative stress is where the rubber hits the road when it comes to the origins of diseases that are related to stress.

A scientific study of illness vulnerability showed that psychological stress in mothers taking care of chronically ill children was associated with ten times the degree of white blood cell telomere shortening, along with low-maintenance enzyme activity and high oxidative stress. In essence, stressed mothers had immune cells that were aging ten times faster than non-stressed mothers.

However, something can be done to halt this rate of decline in our white blood cells' health. Mindfulness meditation appears to reduce oxidative stress and buffer against these aging effects. Meditating can elicit a physiological state called the "relaxation response," which represents a state opposite to the stress response. By doing so, cells become less overworked and the damage is diminished. In experienced meditators, when we look at the genes that are activated in their white blood cells and the ones that are deactivated, we see that those genes associated with the innate immediate inflammatory response—the immune response that quickly begins to battle a microbial or traumatic threat—and those

MINDFULNESS MEDITATION
The process of meditation can begin with focused awareness, often of our breath. It can then proceed to open awareness of thoughts, feelings, and sensations (like the wind and sun in this picture) in a mindful, non-judgmental way, and on to ethical and compassionate reflections. In all there is a reversal of stress and a return to a relaxed sense of wellness.

responsible for cellular aging and oxidative stress are relatively deactivated, while those required for efficient cellular processing of oxygen and glucose are activated. We will explore this and other research on mind–body techniques such as meditation in Chapter Ten.

HOW STRESS AFFECTS THE BRAIN
What does chronic stress do?

Psychological and physical stressors are processed in several of our brain structures including the limbic (for example, amygdala, hippocampus), paralimbic (for example, anterior cingulate cortex), and cortical (for example, prefrontal cortex) areas of our brain. This triangle of the amygdala, the hippocampus, and the medial prefrontal cortex including the anterior cingulate cortex is of key importance in the stress response and how it is modulated.

Regions of the brain
The limbic regions lie deep in the middle of the brain while the cortex is the superficial ribboning of the brain. The paralimbic regions lie between the limbic and cortical areas. Sensory experiences associated with stressors will flow into these areas from lower zones under the cortex such as the thalamus (hearing, seeing, and touching) and olfactory (smelling) nuclei. The processing of our sense's inputs in a more thorough way will come from cortical areas such as the primary sensory cortex, the piriform cortex, and the insular cortex.

Our memory and the context for that memory is accessed from associated areas through the hippocampus and the neighboring entorhinal cortex, as well as other parts of the limbic system, such as the septum, and the cingulate cortex, which is part of the paralimbic region.

Attentional and arousal inputs are strengthened by neurotransmitters added from midbrain regions such as the locus coeruleus (norepinephrine), the ventral tegmentum (dopamine), and the raphe nuclei (serotonin). The outputs from all of these limbic, paralimbic and prefrontal zones converge on what are called "relay sites" in the hypothalamus below the cortex. When this happens, the downstream processing of stress-related information is enhanced. These regions all work together to activate the hypothalamus, the pituitary gland, and the adrenal gland, known as the HPA axis, as well as autonomic and inflammatory responses to stress.

When our amygdala senses danger or a stressor, it stimulates release of the excitatory transmitter called "glutamate" throughout our brain to prepare us for action. This vigilant state of readiness takes up a great deal of our energy. If the state persists or overwhelms us, it will lead to distress, wear and tear, and illness vulnerability. Perhaps the best example of this is what happens in those of us who experience post-traumatic stress. Experiencing traumatic stress intensifies amygdala activity. The overactive amygdala challenges those areas of our brain charged with carrying out stability.

Chronic stress
Chronic stress physically alters the structure and function of brain regions that help control the stress response systems. Severe, unrelenting stress will lead to alterations in the structure of cells in the hippocampus and prefrontal cortex, most notably atrophy of the tiny information-conducting cell spines called dendrites on key memory neurons called pyramidal cells. Chronic stress also causes changes in the hypothalamus, which results in the increased production of the stress hormone called corticotrophin-releasing hormone (CRH), and reduced expression of the receptor for the glucocorticoid cortisol. This receptor is called the GR receptor and it is associated with resilience against stress.

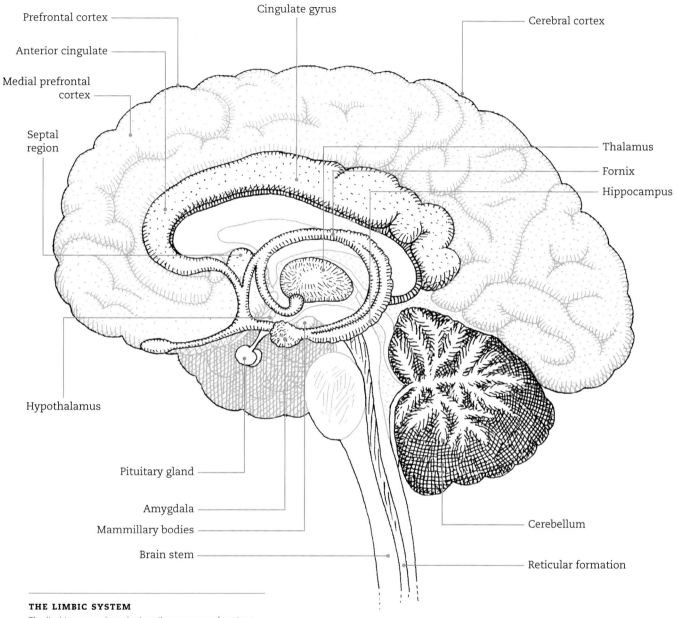

Prefrontal cortex

Anterior cingulate

Medial prefrontal cortex

Septal region

Cingulate gyrus

Cerebral cortex

Thalamus

Fornix

Hippocampus

Hypothalamus

Pituitary gland

Amygdala

Mammillary bodies

Brain stem

Cerebellum

Reticular formation

THE LIMBIC SYSTEM

The limbic system loosely describes a group of ancient brain structures designed to process emotional information and to make it available to the anterior cingulate cortex; this is used for making complicated decisions under stress. It consists of the hippocampus, fornix, septum, mammillary bodies, and amygdala.

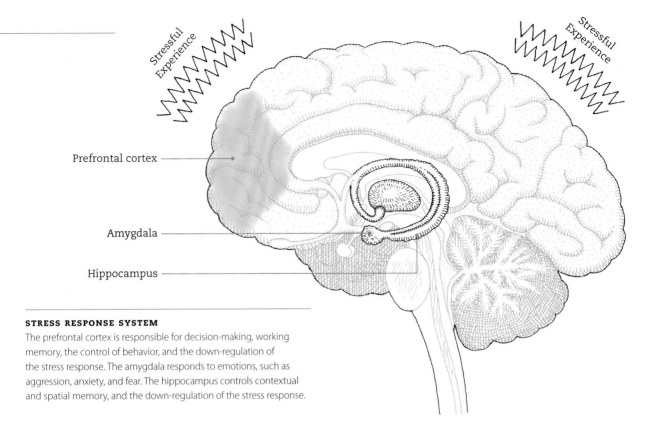

Prefrontal cortex

Amygdala

Hippocampus

Stressful Experience

Stressful Experience

STRESS RESPONSE SYSTEM
The prefrontal cortex is responsible for decision-making, working memory, the control of behavior, and the down-regulation of the stress response. The amygdala responds to emotions, such as aggression, anxiety, and fear. The hippocampus controls contextual and spatial memory, and the down-regulation of the stress response.

The hippocampus

Regions of the hippocampus are known to regulate the HPA axis in stressor-specific ways. For example, lesions of a particular region of the hippocampus can result in increased cortisol release following emotional stress. This reflects how the hippocampus is responsible for context (spatial and temporal) related control of stress responses. The hippocampus also moderates the autonomic nervous system. When it is stimulated, heart rate, blood pressure, and respiratory rate all decrease.

The medial prefrontal cortex can modify how the hippocampus affects stress activity. This cortex is comprised of different subregions that contribute to specific components of the stress response. The medial prefrontal cortex contributes most to HPA axis inhibition in reaction to psychological stressors. This is important in terminating the stress response and regulates glucocorticoid secretion (such as the hippocampus). This means that the medial prefrontal cortex is important for inhibiting stress responses, such as heart rate and blood pressure.

Cells in both the hippocampus and the medial prefrontal cortex are equipped with receptors for glucocorticoid stress hormones. There are two types, called glucocorticoid and mineralocorticoid. Our hippocampus has many of both, while our medial prefrontal cortex has mostly glucocorticoid receptors. This allows a degree of feedback when our cortisol levels rise. The glucocorticoid that resides in both the prefrontal cortex and the hippocampus can serve to dampen the stress response brought about by high levels of circulating cortisol.

Thus, the medial part of the prefrontal cortex and the hippocampus provides a degree of top-down control of the amygdala-driven stress response. In this way, this part of our brain may be the primary arbiter of the physiology of stress. In the face of separation threat, be it a threat to ourself, our family, our home, or our livelihood, our amygdala will take charge and alarm our stress responses. On the other hand, under conditions of attachment-based security, the medial prefrontal cortex makes a more rewarding and calm state available to us.

STRESS AND EMOTION
Why are threats stressful?

We find things most meaningful when we are preserving and enriching the lives of ourselves and other people. Human beings consider threats to themselves and species preservation stressful. When we perceive a threat, for example an unexplained sound in the middle of the night, our sensory brain area (the thalamus) rapidly sends information about the sound as an alarm, using only one synapse, directly to the amygdala for a quick evaluation. This often leads to an acute stress response with all the ingredients mentioned earlier, which prepares us for action. But this quick start alarm mechanism often results in what we call a false positive—it was just the sound of a precariously placed valise a family member used on a recent trip,

plummeting to the floor. This more sophisticated understanding of the sound awaits the cognitive and emotional appraisal of the threat that arises after the transmission of sensory information to other brain regions, including the sensory cortex, certain prefrontal cortex regions, the hippocampus, and the amygdala.

The outcome of this analysis leads to an "at ease" relaxation response, because we realize that the sound does not signify a threat. We do not waste energy remaining fearful and stressed, but instead return to a state of security. But it is good to initially spend some energy getting prepared, because this type of preparation could have saved lives.

FEAR RESPONSE

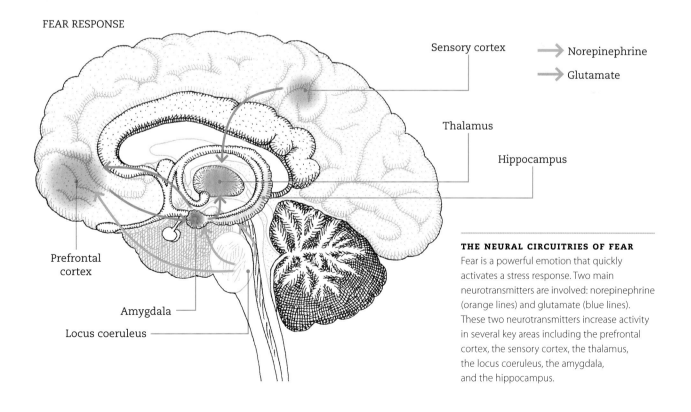

Sensory cortex

Norepinephrine

Glutamate

Thalamus

Hippocampus

Prefrontal cortex

Amygdala

Locus coeruleus

THE NEURAL CIRCUITRIES OF FEAR
Fear is a powerful emotion that quickly activates a stress response. Two main neurotransmitters are involved: norepinephrine (orange lines) and glutamate (blue lines). These two neurotransmitters increase activity in several key areas including the prefrontal cortex, the sensory cortex, the thalamus, the locus coeruleus, the amygdala, and the hippocampus.

Anterior cingulate cortex

There are areas of our brain that evolved to specifically secure our social attachments. Chief among these is the extremely important anterior cingulate cortex, which is part of the paralimbic medial prefrontal cortex. This region helps us when making decisions about whether to avoid or approach anyone or anything.

The anterior cingulate paralimbic cortex mediates what the limbic system researcher, Paul MacLean calls, the "mammalian behavioral triad." This triad includes maternal care for the young; the cry of separation exhibited by all infants when in danger of isolation; and the instinct to play as a way of strengthening attachments. Because of our heritage as mammals, with our reliance on nurturance and social support for our survival, separation challenges

THE FACE OF SEPARATION FEAR

A deep fear of separation lies at the center of the mammalian stress response. This is masterfully depicted by Léon Cogniet in *The Massacre of the Innocents* (1824)—a poor mother and her child stare at us with deep limbic foreboding, with hearts pounding and blood pressure surging in anticipation of the possible loss of their dearest attachment.

and threats are the core elements of all human stressors. Negative emotions are really brain and body generated tags that signify for us separation threats or attachment losses. Positive emotions conversely reflect our secure social attachments.

To summarize, the thalamus, amygdala, anterior cingulate cortex, hypothalamus, hippocampus, and medial prefrontal cortex are key components of a circuit that controls our stress responses.

STRESS AND COGNITION
How do we experience pain and pleasure?

Remember that every human being is ultimately focused on making the basic decision to approach or avoid and that the brain evolved as a separate organ expressly for the purpose of making these movement decisions more effectively. In this context, pain, pleasure, and memory evolved as markers that support our most important and meaningful decision-making capabilities.

We think negatively about things in the context of distressing past experiences and in anticipation of distress in the future. Optimistic appraisals, which often build on past positive outcomes, have the power to reduce our negative emotional responses and to reduce our anticipatory stress. This can make the difference between facing an upcoming separation as a threat to avoid versus a challenge to overcome.

Some of us become chronically negative thinkers. When faced with a stressful experience, we may automatically have a tendency to have a particular distressing thought, which often represents a distortion of the real circumstance. This habitual pessimistic thinking of imagining a catastrophe around every corner, fosters a tendency to favor personal failings—ignoring a multitude of other possible explanations—as causes for any unsatisfactory outcome. The literary character who most embodies these characteristics might be Eeyore, the persistently sad donkey in A. A. Milne's book *Winnie the Pooh*.

As expected, when our automatic thinking is negative, the emotional response attendant to it will be one of anxiety, depression, or anger. The good news is that we can be taught to alter these negative thinking traps. By doing so, we can relieve ourselves of the internal drivers of the chronic stress responses, which often result in anxiety and depressive conditions.

THE BRAIN'S RESPONSE TO PAIN

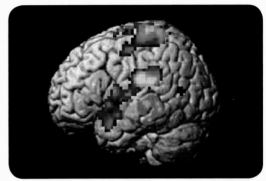

Pain

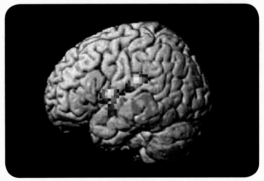

Empathy

EXPERIENCING PAIN
Magnetic resonance imaging (MRI) models show two brains. The top brain is of a person experiencing pain (the active brain areas are colored yellow/red and include the sensory cortex and the cingulate cortex). The bottom brain is of a person watching someone experience pain, producing a response called empathy, where the pain the other person is feeling is understood through the actions of mirror neurons. The MRI scan confirms that some similar brain areas are activated in empathy, but that areas producing the actual sensation of pain (the sensory cortex at the top of the brain) are not triggered.

Cognitive behavior therapy

In a therapy called cognitive behavior therapy (CBT), we are asked to review our automatic reflexive thoughts, and to identify whether they reflect any cognitive distortions. If this turns out to be the case, we can then generate a more adaptive response through exercises that become grist for the mill during therapy sessions over eight to twelve weeks. Eventually, we become capable of changing pessimistic negative thinking habits into more positive ones. Pessimism appears to predispose us to increased stress and wear and tear, leading to negative emotional states. This appears to put us at risk for worse health outcomes, particularly in the area of cardiovascular disease (see Chapter Three).

Stress and memory

When our amygdala reacts in the midst of a threat, memory centers in our hippocampus are stimulated. As a result, the particular memory trace can be strengthened and will have a lasting remembrance. However, when stress becomes chronic or overwhelming, our hippocampus can become overly excited, even leading to structural damage with memory function suffering a blow.

Stress hormones like cortisol influence a wide range of our memory functions. It turns out that cortisol enhances consolidation of memory (the creation of new memories by stabilizing a memory trace after the initial acquisition), but impairs memory retrieval (re-accessing of previously encoded and stored memory traces). In this context, GC receptor sensitivity and hippocampal integrity play important roles. With this in mind, how might this structural damage occur?

When there are unremitting fear responses pouring from the amygdala into the hippocampus, as occurs in post-traumatic stress disorder (PTSD), information will be preserved that is important for survival. At the same time, there is sometimes impairment of memory retrieval and working memory, and stress becomes chronic. There are two damaging effects that can impair our hippocampal functioning. First, the dendrites—small protrusions on the neuron that improve cell-to-cell communication—are less effective by becoming shorter and less branched. Second, neurogenesis—the birth of new nerve cells—in the part of the hippocampus called the dentate gyrus is reduced, making hippocampal memory function more difficult. While these changes are by and large reversible when stress remits, under chronic persistent stress, these damaging effects may become permanent. Indeed in conditions such as PTSD, shrinkage of the hippocampus has been noted on neuroimaging showing the damaging effects on memory function and behavior.

The medial prefrontal cortex normally regulates emotional responses and takes part in working memory and attention. Its functions are impaired under stress conditions

Medial prefrontal cortex

Chronic and severe stress

Hippocampus

Amygdala

The amygdala grows larger, and forms more connections to other neurons. It becomes hyperresponsive, inflating the fear response.

The hippocampus shrinks, as neuron growth slows. Fewer connections to other neurons form. Memory function is impaired.

STRESS-RELATED STRUCTURAL CHANGES IN INTERCONNECTED BRAIN REGIONS

Under conditions of severe stress, the amygdala gets stronger and shows structural and functional evidence for this. Meanwhile, the hippocampus and the medial prefrontal cortex work hard trying to modulate the amygdala and, as a result, they wear down structurally and functionally over time.

STRESS AND AGING
How do our brains age?

There is evidence that stress resulting in allostatic loading can increase the risk for dementia as well as depression in elderly people. This risk may relate to the chronic activation of the immune system that comes with chronic stress (see also Chapter Four). Remarkably we can see this in the way genes are expressed in white blood cells in individuals who are stressed. In stressed-out individuals, cells have a common pattern of gene expression. For example, the genes responsible for promoting inflammation are stimulated along with genes that prevent receptors for insulin from working efficiently. The result is a more rapidly aging cell.

Our brains are particularly vulnerable to the effects of chronic stress. Fewer GC receptors are produced in the hippocampus of the brains of chronically stressed individuals, reducing their ability to attenuate the stress response. As we age, we lose the capacity to manage an abundance of excitatory transmitters. Excessive glutamate-induced excitement forces mitochondria (the cell organelle responsible for producing cellular energy through processing of glucose and oxygen) to work overtime. The result can be damage from byproducts, called reactive oxygen species or free radicals. Imagine too much oxygen processing causing cellular rust build-up. Cell damage can ensue from this excessive oxygen reactivity. The high concentration of these oxygen metabolites in a cell has been shown to accelerate cellular aging and to promote the onset of dementia.

Stress-related cortisol output sensitizes neurons to glutamate and increases the number of glutamate receptors in our brain, making overexcitement and free radical damage more problematic.

Aging also comes with a slowing of neurogenesis; this makes the reduction of new neuron growth that comes with stress more critical to age-related illnesses. And, in what may partly reflect stress-related inflammation, scar producing senile plaques, containing dead microglial cells, accumulate in various forms of dementia.

ALZHEIMER'S DISEASE CAN CAUSE DEMENTIA

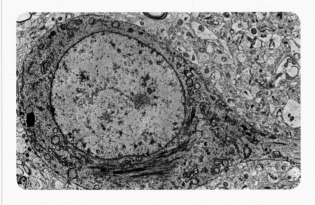

ALZHEIMER'S DISEASE

Alzheimer's disease is the most common cause of dementia. In this colored transmission electron micrograph (TEM) of a nerve cell from the brain of a patient with Alzheimer's disease, tangles of protein (dark blue) can be seen in the cytoplasm (green). These "neurofibrillary tangles" are aggregates of tau proteins. In a healthy cell, these proteins stabilize features called microtubules, which play a number of essential roles in the cell. Oxidative stress can cause the normally soluble tau proteins to form these insoluble tangles, interfering with normal cell functioning. This often results in the cells dying.

NEUROCHEMICAL MESSENGERS
The brain's chemical networking system

Stress affects a variety of neurochemical messengers in our brain and body. Chemical messengers include neurotransmitters, neuropeptide hormones, endogenous opioids, and cytokines. These messengers are also bi-directional with regard to the immune system and the brain. Cytokines affect our brain while neurotransmitters affect our immune system.

Monoamines
The monoamines include norepinephrine, dopamine, and serotonin. These transmitters are increased in specific neuronal populations in our brain soon after the experience of acute stress. Stimulation of these monoamines depends on a variety of factors including gender, time of day, as well as the nature and duration of the stressor. The hippocampus, amygdala, prefrontal cortex, and our brain's reward center, the nucleus accumbens, all receive monoamine input.

Our brain's release of monoamines under conditions of stress occurs within minutes and as a result our reactions will be rapid after these messengers attach to receptors and activate signals to downstream action centers that effect movement. This results in functional changes in our behavior.

Neuropeptides
There are also several neuropeptides that contribute to the stress response. They do this by stimulating multiple receptors in specific brain regions. The neurohormone CRH is a key stress-related molecule but there are other peptides that counteract the stress response (vasopressin, oxytocin, and neuropeptide Y—these will be described in more detail in later chapters).

The hormone CRH is released in response to stress from nerve terminals in the hypothalamus and acts on receptors in the pituitary gland setting in motion

CHEMICAL MESSENGERS

In the context of stress, each monoamine has a particular contribution to make.

Dopamine enhances our risk assessment and risk prediction and helps with our decision-making.

Norepinephrine may help us attend to the surroundings more intently in a search for sensory input that could support better solutions to challenges.

Serotonin soothes the anxiety that comes with a stressor.

Thus, monoamines cooperate in the promotion of our resilience helping us to cope in the face of a significant stressful event.

MONOAMINES

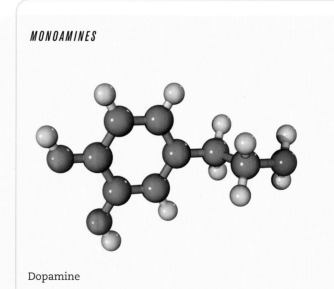

Dopamine

the HPA stress response. CRH is also expressed in parts of the amygdala, the hippocampus, and the locus coeruleus. When CRH is released in the hippocampus in the face of modest stress, memory is improved, but when the hippocampus releases large quantities of CRH in the setting of severe overwhelming stress, hyperexcitability may arise along with loss of dendritic spines in hippocampal cells as described above. Chronic stress also leads to structural changes in which CRH plays a role.

The adrenal glands' secretion of cortisol varies in relation to a daily circadian rhythm characterized by a surge during the morning hours. Secretion also fluctuates in relation to stress factors. Cortisol levels are synchronized in the brain. This allows coordination of peripheral and central aspects of our stress response. As mentioned above, cortisol works primarily through actions at mineralocorticoid (MR) and glucocorticoid receptors (GR).

Certain of our brain regions like the prefrontal cortex, the amygdala, the hippocampus, and the midbrain monoaminergic synthesis areas—the ventral tegmentum (for dopamine), the dorsal raphe nucleus (for serotonin), and the locus coeruleus (for norepinephrine)—are particularly sensitive to stress

TIMING AND COORDINATION OF THE STRESS RESPONSE

The stress response occurs in coordinated phases over varying time courses.

Phase one: Opening acute wave of activities, within seconds of a stressful event; mediated by monoamines and neuropeptides. It is designed to elaborate the stress response physiologically and immunologically.

Phase two: Molecular effects of stress; within one to two hours of stressful event; mediated through glucocorticoid receptors activation. It is designed to slow the long-term negative health effect of allowing the first phase to persist.

Initiation of the acute stress response is crucial, but so is the capacity to shut it down appropriately. This restores stability in the face of change for cognitive, emotional, physiological, and neuroendocrine functions.

mediators and messengers. These areas serve as important hubs that link up different brain networks integral to the stress response. The integration of all these messages enables us to finely tune brain responses to a myriad of stress experiences.

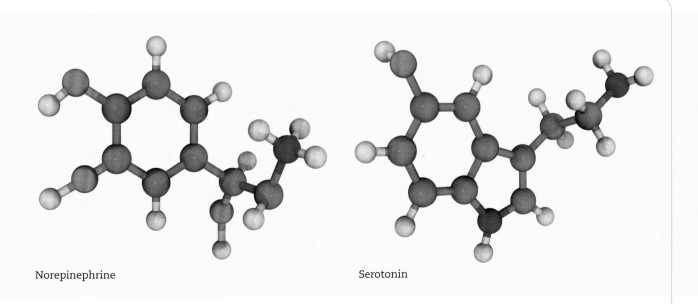

Norepinephrine

Serotonin

STRESS AND ADDICTIONS
Is stress linked to addiction?

Whether we are talking about addictions to drugs or to behaviors, addiction disorders are more likely to arise in the setting of distress. When faced with stress, people will turn to substances or behaviors that will activate their reward circuitry in an effort to bring them pleasure instead of pain.

Addictive disorders, including addiction to alcohol, opiates, or cocaine, come with tremendous human suffering as well as a huge public health price tag. This makes examination of the role of stress in initiation, craving, and relapse an important area of study. Addictions can be conceptualized as chronic relapsing brain diseases. Normal brain mechanisms that give us a sense of reward and pleasure become distorted in the face of more powerful substance-induced direct effects. This can lead to destructive persistent changes that occur at molecular, cellular, behavioral, and psychological levels.

These changes (or neuroadaptations) are predicated on specific drug effects at the receptor level. They are related to three factors. The first relates to the environment, which includes stress-inducing circumstances. Research reveals important interactions between the stress-responsive systems and our vulnerability to abuse disorders. The common stress-related cellular gene expression state mentioned above is characterized by cellular strain from oxidative stress and a heightened inflammatory response. This state may increase vulnerability to addiction. Remarkably, addictions to specific drugs such as alcohol, stimulants (for example, cocaine and amphetamines), sedative-hypnotics (for example, benzodiazepines), and opiates (for example, heroin) all have a common effect on the brain. The neurotransmitter dopamine plays a key role in this effect. Our brain contains a reward circuitry that rewards us with a pleasurable feeling whenever dopamine functioning increases in a brain nucleus called the nucleus accumbens. All drugs of abuse do precisely this. They increase dopamine functioning in the nucleus accumbens. And by doing so we always take the risk of having our precious brain reward system being hijacked by the artificial high of recreational drugs, with the possibility of never again experiencing the normal pleasures of life. The saddest example of this is the loss of the sense of parental reward in the simple act of caring for one's child.

Drugs of abuse have common direct or downstream effects on several brain stress-responsive systems, including those for oxytocin, vasopressin, and for the body's own opioids. Further understanding of these stress-responsive systems may help us find early interventions and to discover new treatments for addiction disorders.

Acute alcohol ingestion leads to stimulation of the stress systems, particularly the HPA axis. Those of us who chronically ingest large amounts of alcohol develop a tolerance to this stimulant effect of alcohol. In other words, the stress systems get used to stimulation after alcohol. But the brain pays a heavy price for this effect in the long run.

Through its effects on our brain, alcohol can lead to gait problems, blurry vision, slurred speech, slowed reaction times, and decreased memory functioning. While these effects will dissipate when drinking ceases, when drinking is persistent and heavy, brain deficits can continue well after the person stops drinking. Research into how alcohol itself as a toxin might cause long-term memory deficits and even a dementia is ongoing. Certainly it is an established fact that nutritional deficiencies that accompany alcoholism can cause brain damage. Thiamine deficiency, for example, causes specific damage to the brain leading to memory disorders called the Wernicke-Korsakoff Syndrome.

Several factors influence how alcohol affects the brain. These include:

▶ The quantity and frequency of the alcohol consumption;
▶ The age when drinking began, and how long drinking has continued;
▶ The age, gender, genetic background, educational level, and family history of alcoholism;
▶ The history of possible fetal alcohol syndrome from prenatal alcohol exposure; and
▶ The general health status.

When alcoholic individuals abruptly stop drinking, a host of other potentially severe medical consequences including seizures and so-called delirium tremens may ensue. These conditions can be life threatening.

Addiction to substances wreaks havoc with the human stress response system. Like the acute stimulatory effect of alcohol, acute cocaine use stimulates HPA activity in human beings. The chronic action of drugs of abuse promote compensatory changes in the dopamine reward circuitry that resemble chronic stress. Drugs of abuse hijack the usual brain reward signals, and this means that behaviors that are normally felt to be calming and pleasurable no longer have the same effect or potency on the individual.

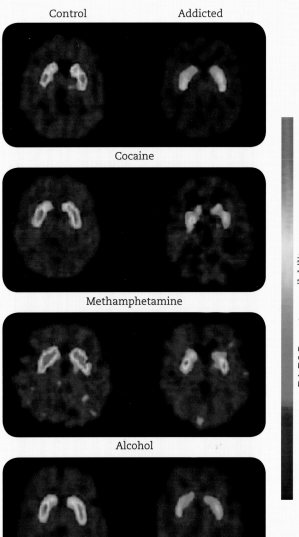

THE EFFECTS OF SUBSTANCE ABUSE

BRAIN DAMAGE CAUSED BY DRUGS AND ALCOHOL
The Dopamine D2 receptor has been implicated in drug abuse as well as vulnerability to become addicted. Chronic drug use reduces the function of D2 receptors. In addition, those who genetically start out with fewer D2 receptors in the brain are more likely to be impulsive, and more likely to become addicted to alcohol and drugs. These images show reduced D2 receptor function in the medial prefrontal regions in a PET scan of addicted subjects as compared to controls.

ADDICTED TO EXCITEMENT
The thrill of stress

ndividuals may become "addicted" to the excitement that comes with a challenge. These individuals have been called "adrenaline junkies." They are thrill seekers who get pleasure out of the rush that comes with a struggle or threat. They ski down huge mountains, bungee jump, or surf huge waves. The world-famous rock climber and base jumper Dean Potter died age 43 in the course of a base jump flight in Yosemite. He was on a crusade to make these death-defying jumps legal in the US. Some of us of course plug into this energy vicariously by sitting on the edge of our seats while enjoying extreme sports shows and while watching action or horror films.

As mentioned, the nucleus accumbens is the key brain region concerned with pleasure and reward. It is a central region where emotionally derived motivations interface with our motor circuitry to help us achieve our goals. Because of this the nucleus accumbens becomes central to the addictive process in general and plays a role in goal-directed movements. This is because the brain reward system can be hijacked not only by substances but by pleasurable behaviors that abnormally accentuate dopamine receptor function, artificially rendering normal pleasures such as family connection, play, and work activity rather mundane and relatively pleasureless. The thrill of a stressful dangerous challenge sometimes bordering on recklessness or hypersexuality can appropriate the same reinforcing brain reward circuitry and become a habit.

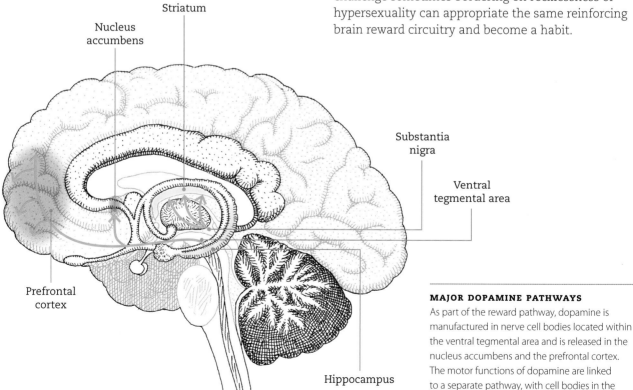

Striatum

Nucleus
accumbens

Substantia
nigra

Ventral
tegmental area

Prefrontal
cortex

Hippocampus

MAJOR DOPAMINE PATHWAYS
As part of the reward pathway, dopamine is manufactured in nerve cell bodies located within the ventral tegmental area and is released in the nucleus accumbens and the prefrontal cortex. The motor functions of dopamine are linked to a separate pathway, with cell bodies in the substantia nigra that manufacture and release dopamine into the striatum.

There are some of us who may be at higher risk for being thrill seekers. Neuroimaging studies have shown that thrill-seekers and risk takers have fewer inhibitory dopamine-regulating receptors than those of us who are more shy and retiring.

This suggests that those of us who crave excitement through taking risk may experience a large burst of dopamine each time we have an exciting experience. This may relate to the fact that risk-taking brains are less able to inhibit the neurotransmitter dopamine adequately. And another study suggests that when completing a learning task, risk-taking individuals

STRESSFUL CHALLENGES

Challenging situations can be rewarding. Some of us will engineer a stressful episode in order to recreate the prototypical stress of a threat to one's attachments. The pay-off is the powerful sense of gratification this ice climber will likely feel at returning to equipoise and security.

with fewer dopamine regulatory receptors experience a weaker negative feedback. So those of us who are risk takers may crave excitement and take extreme risks because we have a harder time learning from our mistakes and negative experiences.

POSITIVE STRESS
When is stress beneficial?

Stress can focus our attention and help us to achieve our goals and overcome our obstacles. The stress response can be associated with the arousal experiences of athletes locked in the heat of competition or the exhilaration of a parachutist at the moment of leaping from the doorway of a plane. It is also true that the joyous excitement of meeting a stressful challenge contrasts with the distress of facing a dreadful threat. The former contributes to our zest for life and often leads to leaps in our personal development. Distress, by contrast, is associated with a fear of failure and the unfulfilled wish for achievement of a desired end, resulting in a feeling of deprivation.

Distress can eventually lead us to failure and frustration in the meeting of stressful challenges. This sense of failure can lead to a condition known as "learned helplessness," which carries with it all the hallmarks of what may be thought of as a chronic stress response, namely depression. When any of us is subjected to random inescapable stressors, chances are we will enter a state of hopelessness and helplessness associated with high cortisol output. We can begin to show a petering out of our fight-flight responses in favor of what has been called a "giving-up, given-up state." Some theorists believe that the tendency to adopt this passive defensive strategy of so-called "conservation-withdrawal" in the face of inescapable stress, may stem from an evolutionary drive to save energy. When we suffer the chronic stress-related condition of major depression, we experience a similar symptom cluster, namely, poor sleep; loss of interest and pleasure in usual activities; hopelessness, helplessness and worthlessness; loss of energy; loss of concentration; loss of appetite and weight loss; psychomotor agitation or retardation; and morbid or suicidal thinking.

Whether we appraise a stress as threat or as challenge appears to correlate with the potential outcomes of learned helplessness versus what has been called "learned optimism." When we can summon up the courage to face our stressors with optimism, purpose, and meaning, we can increase our chances of successful adaptation and of developing resilience, and of even experiencing what has come to be known as post-traumatic growth. In this way, stress can indeed be a positive experience.

STRESS ENERGIZES US
Leaving the safety of an airplane and leaping into uncertainty sets our stress response systems into high gear. Our brains and bodies will prepare us to meet the threat.

The performance edge

While stress is often thought of as negative, it is important to realize it can serve positive ends as well. Changes and challenges that stress us can mobilize us to achieve our goals and help us to learn and mature. Managing stress is to the brain what building muscle through exercise is to the body. Of course finding the right balance between pumping too much or too little weight and running too little or too much is key. So it is with stress: not enough stress leads to boredom and lethargy; too much stress leads to reduced performance and damage.

But there is also another secret to high performance under stress. This is to practice enough—remember the 10,000-hour rule for mastery of any skill or subject area—to be able to place yourself in a comfort zone aided by the physiological state of relaxation response to buffer against an excessive stress response. This has been called "being in the zone," a place of high performance all athletes aim for when they are in competition and which the military tries to instill in their soldiers. So serving at match point or learning to target shoot under pressure would be good examples. Of course the same can be said of performance in any sphere— singing opera, playing chess, delivering a lecture in front of a large audience and so on.

Learning how to use stress to our best advantage—to energize us as opposed to exhausting us—is a boost to success in any field. And understanding how stress is managed in our brain is essential to promoting both health and performance. We have gotten a good start in our understanding of this challenging area. There will be several opportunities in the following chapters to revisit a number of the key concepts introduced in these first two chapters.

STRESS ENHANCES PERFORMANCE
Performing in front of others in any vocation is accompanied by a certain amount of stress, which can sharpen our skills and energize us to reach new heights. For example, experienced singers can harness stress to create outstanding and remarkable performances.

Chapter Three

STRESS AND THE CARDIOVASCULAR SYSTEM

Your brain and your heart are intimately connected organs. Your brain is charged with decision-making to keep you safe by mobilizing or immobilizing you. To do this, it needs to be in constant contact with the heart in order to manage the heart rate and blood pressure responses necessary to supply blood and nutrients to your skeletal muscles, in the context of the stress response, and to your smooth muscle and viscera, when you are at rest. Your brain itself also needs this support to carry out your thinking functions.

THE NERVOUS SYSTEM AND THE HEART
The risk of heart attacks

When your brain becomes tired through the stress-related metabolic wear and tear of allostatic loading, there will be a tendency for you to be more vulnerable to a variety of illnesses. It should not be a surprise that the heart is especially vulnerable in the face of stress. Indeed evidence suggests that stress and stress-related emotional disorders such as anxiety and depression are associated with increased risk for cardiac conditions, such as heart attacks (myocardial infarctions in which heart muscle is damaged) and cardiac arrests secondary to abnormal heart rhythms (cardiac arrhythmias that may end in heart stoppage). An over-reactive and chronic stress response dominated by the sympathetic nervous system can also foster the development of heart failure, in which the chambers of the heart inadequately push blood to the body's tissues.

Intriguingly, acute coronary events can even take place in those of us who lack the traditional risk factors for coronary artery disease (CAD). This highlights the importance of stress-related factors in our lives, those challenging psychological and social situations we all face on a daily basis.

Stress and the heart

Chronic stress can predispose us to cardiac ischemia—the reduction of blood flow to the heart that results in cardiac muscle death—and cardiac electrophysiological accidents—the aberrant electrical impulses in the walls of our heart's chambers that lead to poorly-timed contractions and inefficient functioning. These events damage the heart and potentially lead to death. It has been reported that up to half of patients with coronary heart disease will experience transient, painless "silent" myocardial ischemia during mental stress. Silent myocardial ischemia can be very dangerous because patients do not feel angina (heart pain) even when their heart muscle is being deprived of blood. This can leave unfortunate individuals with silent ischemia without any pain warning that their heart muscle is at grave risk.

Disturbingly, many ischemic episodes detected by ambulatory electrocardiograms in chronically ill cardiac patients are silent and not caused by physical exertion. Indeed, these episodes are usually associated with mental stress. This syndrome of what has been termed "mental stress ischemia" contributes to a three-fold increase in the risk of poor clinical outcomes in heart patients. Simply put, stress-related heart disease contributes mightily to your overall cardiac risk.

Taken together with the high prevalence (presence of the particular disorder over a lifetime) of heart disease in the general population and the high incidence of new cases of heart attacks during man-made and natural disasters, these findings should prompt all of us to improve our ability to manage stress in our lives and in our communities. We should also help to identify those who are more vulnerable to stress-related heart disease, so we can help them find ways to reduce their stress.

CARDIOVASCULAR DISEASE (CVD) RISK FACTOR RESPONSE TO MENTAL STRESS
(Response after three mental stress tasks: mental arithmetic, mirror trace, anger recall)

CARDIOVASCULAR REACTIVITY Alterations in coronary artery response to stress	MENTAL STRESS-RELATED ISCHEMIA Mental challenge induced reduction of blood flow to tissue	PLATELET AGGREGATION More risk of thrombus (clot) under stress	PSYCHOSOCIAL STRESS Anxiety, depression, and negativity are connected with cardiac risk
Men show larger changes in physiological markers, such as blood pressure than women	Women displayed more mental stress-induced myocardial ischemia (MSIMI) than men	Women have a greater increase in platelet aggregation under stress conditions than men	Women showed more negative emotion, and less positive emotion than men

Outcomes and CVD prognosis for women are worse after prognosis, compared to men

STRESS, GENDER, AND CARDIOVASCULAR REACTIVITY
While men appear to respond more to stress at the blood vessel level, women respond more vigorously to mental and psychosocial stress and have greater platelet viscosity effects leading to particular cardiac vulnerabilities in a stressful environment.

The central autonomic network

Why might stress-induced heart disease occur with such frequency? Researching brain activity related to stress has provided us with some hints. Several areas of our brain are known to be associated with our comprehension of stress and several areas activate in response to distressing conditions. Included in these brain territories are the regions that process our emotions and the cognitions stimulated by our experiences.

Within our brain there exists a "central autonomic network." This is a dynamic system made up of connected brain zones cooperating to process similar stress-related activities involving our heart. In essence, this creates a structural brain–heart connection. This central autonomic network is thought to modulate our stress-related emotional responses. The nerve connections within the network regulate and integrate stress responses in our autonomic nervous system through a series of feedback mechanisms designed to maintain our allostasis.

Key regions in this brain–heart connection include prefrontal regions we have encountered before, such as the medial prefrontal cortex and the anterior cingulate cortex discussed in Chapter Two, as well as the insular cortex. The amygdala of course also contributes to brain–heart communication.

The medial prefrontal cortex is known to be involved in processing the emotional tone of what we experience. It evaluates the significance of an emotion and has the capacity to dampen any over-response we might have to stress. Structurally, the anterior cingulate cortex has access to memory from the hippocampus and future planning from another part of the prefrontal region. These relationships enable this brain region to help us choose proper responses, especially when there is conflicting information for us to consider.

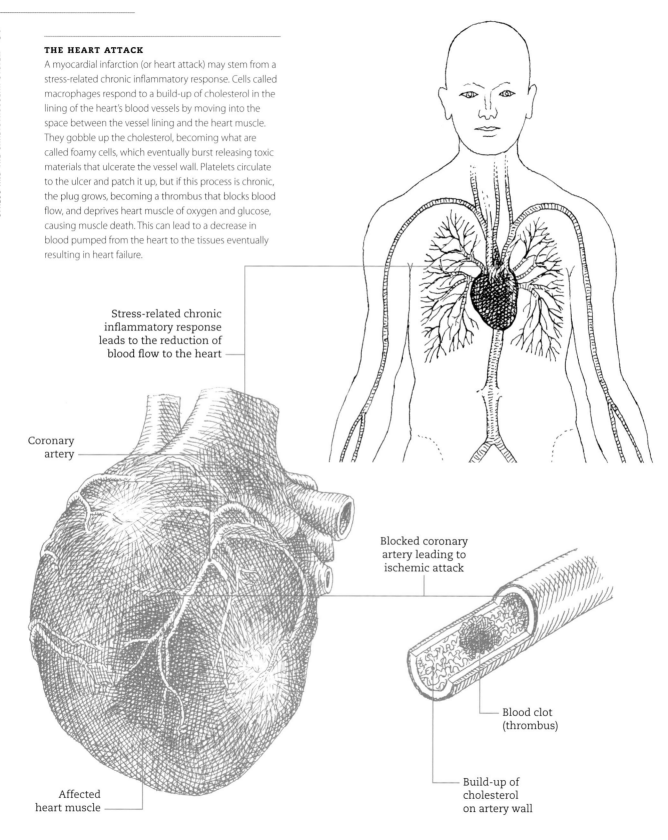

THE HEART ATTACK

A myocardial infarction (or heart attack) may stem from a stress-related chronic inflammatory response. Cells called macrophages respond to a build-up of cholesterol in the lining of the heart's blood vessels by moving into the space between the vessel lining and the heart muscle. They gobble up the cholesterol, becoming what are called foamy cells, which eventually burst releasing toxic materials that ulcerate the vessel wall. Platelets circulate to the ulcer and patch it up, but if this process is chronic, the plug grows, becoming a thrombus that blocks blood flow, and deprives heart muscle of oxygen and glucose, causing muscle death. This can lead to a decrease in blood pumped from the heart to the tissues eventually resulting in heart failure.

Stress-related chronic inflammatory response leads to the reduction of blood flow to the heart

Coronary artery

Blocked coronary artery leading to ischemic attack

Blood clot (thrombus)

Affected heart muscle

Build-up of cholesterol on artery wall

The anterior cingulate cortex mediates all of our pain signaling, including heart-related angina pain along with other of our bodily and emotional pain states. This structural arrangement in the anterior cingulate cortex hub makes perfect evolutionary sense. We would want to combine inputs from memory, planning, and pain signaling, before making any of life's decisions.

The amygdala has been shown to be critical to the processing of fear, and also contributes to the stress-prone disposition of certain individuals. The amygdala will be engaged in looking out for threat-related stressors and sets the tone for the pathways that results in activation of the stress response system.

Neuroimaging studies have revealed that the network consisting of the medial prefrontal cortex, the anterior cingulate cortex, and amygdala, as well as a part of the brain called the insular cortex, are integral to the regulation of the brain–heart connection. The insular cortices play a large role in connecting the brain to the body.

The effects of having a stroke

Damage to the insular cortex, as can occur in a stroke, may adversely affect cardiac prognosis especially when the left insular region is damaged. The culprit appears to be an imbalance between parasympathetic (vagus nerve) and sympathetic nervous system activity. This is especially damaging in the presence of co-existing coronary artery heart disease. Psychosocial distress can be a risk factor, since it can strongly activate your sympathetic nervous system, worsening the imbalance and lowering the threshold for heart rhythm disturbances, heart muscle damage, and heart failure.

In addition, chronic stress can increase cardiac risk by aggravating processes that lead to blood vessel inflammation and deposition of cellular materials that cause blockages in the vessels themselves.

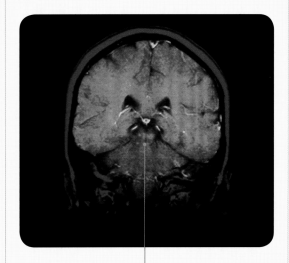

An axial view showing the reduction of blood supply

A coronal view through the normal human brain

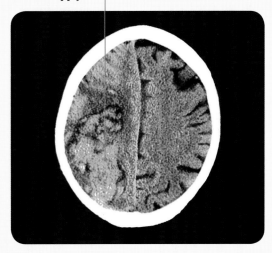

MRI OF A BRAIN AFTER A STROKE
Magnetic resonance imaging (MRI) scans of sections through the brain of a patient following a stroke. A stroke, or cerebrovascular accident, is the rapid loss of brain function due to structural changes from disturbance in the blood supply. Thus an "ischemic" stroke is caused by a reduction in the blood supply to an area of the brain. Chronic stress can increase the risk for such a stroke.

STRESS AND THE NEUROLOGY OF CURBING ACTIVITY

How the body reacts to life's challenges

The intense stressors we face such as loss of a job, or injury to a child, or marital discord will activate our brain–heart connection. This activation will be strongly associated with the emotional area deep in the brain called the limbic system. The amygdala is a major player in the limbic system. When the limbic system is excited by a threat or challenge and mounts a stress response, the medial prefrontal cortex will tend to regulate the amplitude and duration of the stress responses. This can be thought of as a negative feedback circuit between the limbic system and the prefrontal cortex. When the limbic tone gets too high this cortex will dampen it, like the adjustment that occurs with a sophisticated thermostat. However, if our state of stress becomes prolonged or is out of proportion to the trigger, it is likely that situational imbalance exists between prefrontal and limbic activity, or there is a fixed inability of the cortex to diminish our limbic responses. If this were to happen, we would find ourselves uncomfortable, anxious, and perhaps fearful much of the time.

Autonomic nervous system dysregulation may play a role here. Normally, there is a balance between sympathetic activation and parasympathetic deactivation. Emotional distress can result in a net shift toward a fight-or-flight response with increased sympathetic acceleration tone and/or withdrawal of parasympathetic vagal braking tone. If this sympathetic overdrive strikes you, it may place your heart at risk for cardiac arrhythmias.

In functional neuroimaging experiments, patients with heart disease show greater activation of the connections from the limbic system to the prefrontal cortex. This reflects the fact that an overactivated amygdala, as mentioned, is responsible for fear and anxiety responses that can produce a strain on cardiovascular function. Among coronary heart disease patients, those with stress-induced silent heart disease, without cardiac pain, have even greater activation in the medial prefrontal cortex when compared to those who have non stress-related heart disease. This may reflect the exertion in this cortex required to reduce excitation flowing up from the limbic amygdala.

Resilience

If you can weather the storm of a stressful event or phase in your life, you are said to be resilient. Resilience means you can make good adjustments across different domains in the face of significant adversity. It describes the capacity of a dynamic system to withstand challenges to its stability, viability, or development. This will be reflected in better energy dynamics—reducing expenditure and building reserve.

This resilience is characterized by the capacity we have to dampen the stress-induced firing that takes place in the deep emotional centers in our brains, particularly in the amygdala, which leads to the cascade of stress response changes described above. Prefrontal regions of the brain–heart connection do the job of modulating the amygdala, when the system is working in a healthy manner.

It is instructive for us here to consider the special case of those who unfortunately suffer greatly after having experienced a severe traumatic event. We can surmise that the severe stress of post-traumatic stress disorder (PTSD) will eventually lower activation levels in the medial prefrontal cortex region, as can be tracked in brain scans, in a person with PTSD. This suggests that over time, as the capacity of this cortex to control amygdala-driven stress weakens from overuse and then wanes, the toll taken on the heart will increase in line with a persistent stress state. There is evidence that this may be the case. Army veterans with late-life PTSD appear to be at a 45 percent increased risk for heart disease and a 49 percent increased risk for myocardial infarction compared with veterans without late-life PTSD.

THE EFFECTS OF HEART PAIN ON THE BRAIN

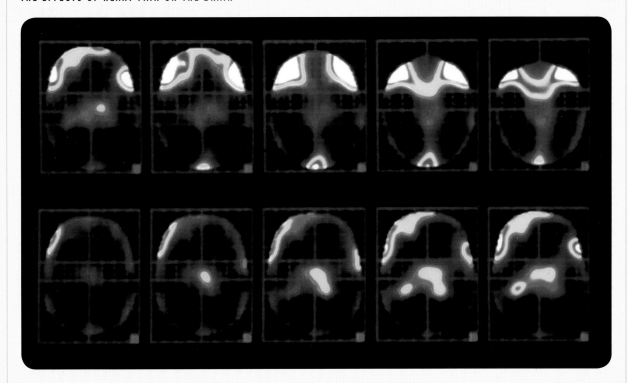

PAIN DURING ANGINA

Scans of the brain of a patient during an attack of chest pain secondary to reduced blood flow in blocked coronary arteries. This process is known as angina pectoris. The brain is viewed in horizontal section, with the front of the head facing up. Blood flow is color-coded: white, red to yellow (high); green, blue to black (low). The most highly activated brain regions have the highest blood flow. The increased brain activation depicted reflects the experience of heart pain. In the bottom row, the thalamus (the bilateral sensory nuclei at the brain's center) is active and it feeds forward sensory information into the brain. In the top row, frontal cortical regions including the anterior cingulate cortex and the insular cortex are activated as pain is consciously being experienced. Studies have shown that under stress, patients with heart disease had significantly increased blood flow to the front cortical regions, especially areas referred to as paralimbic regions that synthesize emotional and cognition information while processing pain signals. These areas are connected with emotion, memory, and stress regulation. This suggests that brain regions that modulate fear and anxiety can affect heart function. The brains of patients without heart disease were more active in cognitive, rather than emotional, centers of the brain.

Type A and type B personalities

Your personality traits are behavioral dimensions that reflect how you interrelate with other people and with society at large. These traits are thought to derive partly from your genetic inheritance and partly from early life experiences, which probably interact with the tendency for some of your genes to become over- or underactive. This will affect your protein concentration, structure and function, and eventually your behavioral tendencies. Many theorists describe five personality dimensions associated with the so-called "emotional IQ," see below.

TYPE A PERSONALITY

Conscientiousness characterized by reliability, carefulness, and persistence

Negative emotionality with traits such as proneness to anxiety, irritability, sadness, and insecurity

TYPE B PERSONALITY

Agreeableness reflected in regard for others, empathy, and generosity

Positive emotionality with traits such as sociability and energy

Openness to experience marked by creativity, imagination, and insight

Personality traits can overlap and moderately predict a range of health and psychosocial outcomes.

The concept of the stressed-out type A personality emerged almost 50 years ago and became associated with a risk of coronary artery disease. The more relaxed type B personality was sketched out in contrast to the type A, in that type Bs showed a positive emotionality and agreeableness. In the schema above, type As may be said to display more negative emotionality along with conscientiousness. The type A personality traditionally was described as a composite of hostility, time urgency, competitiveness, and dominance.

More recent work has focused attention on the personality trait of negative emotionality and its link to cardiac disability and death. Whether anger, anxiety, and depression are distinct emotional

states, or whether they reflect a general tendency to negativity, is still an open question. Regardless, it is now accepted that depression alone is an independent risk factor for coronary artery disease and for worse outcomes after a cardiac event. This simply means that if you suffer from depression your chances of developing heart disease are increased, and your chances of doing poorly after a heart attack are also increased.

Depression may affect your heart in a variety of ways given the fact that heart disease results from a chronic stress response. Thus we can see depression predisposing to metabolic syndrome as described in Chapter One (hypertension, high lipids, poor insulin responsiveness, and obesity) and its tendency to cause cardiac disease. A typical cardiac patient is someone with obesity, high blood pressure, high cholesterol, and type 2 diabetes. And this profile can start out with chronic stress that coincidentally also predisposes to depression. Depression also leads to dangerous behaviors such as sedentary lifestyles, smoking, and non-compliance with medication and clinical advice.

Type D personality

Type D or the distressed personality type, first discussed in the 1990s, has been proposed to combine the cardiac risk factors of depression and poor social support. Negative emotionality and shyness or social inhibitions are key underlying factors and it is characterized by stress-related feelings of negativity, depression, anxiety, anger, and loneliness. The type D personality has been shown to predict poor clinical outcomes in cardiac patients. These personalities "sweat the small stuff" and "make mountains out of molehills" (catastrophize) often expecting the worst possible thing to happen. This creates a great challenge for these individuals after they have developed a cardiac problem, because they can become consumed with the fear of recurrent cardiac illness and as a result risk becoming what are called "cardiac cripples." They also think they are unworthy and find it difficult to make close friends, so their social network of support may be small. They are "tightly wound" and tend to be "hot-reactors" in stressful situations.

THE INTERRELATION OF DEPRESSION AND HEART DISEASE

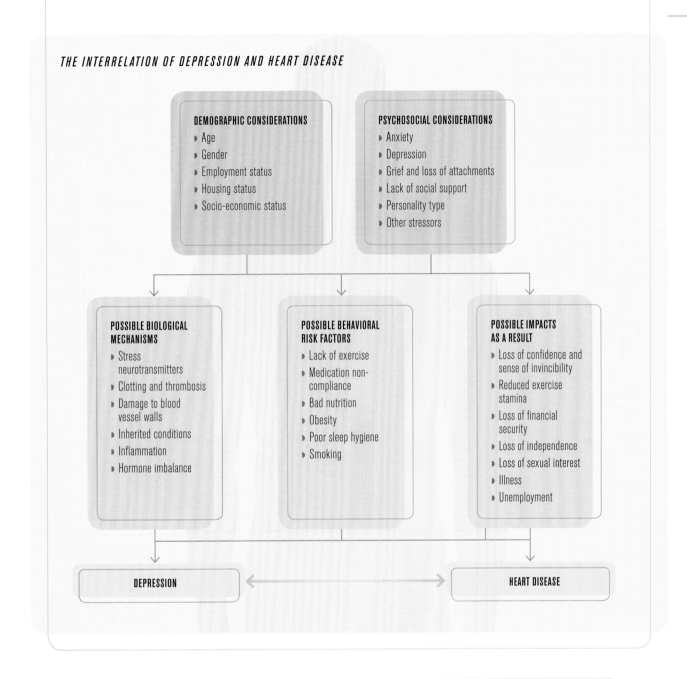

DEMOGRAPHIC CONSIDERATIONS
▸ Age
▸ Gender
▸ Employment status
▸ Housing status
▸ Socio-economic status

PSYCHOSOCIAL CONSIDERATIONS
▸ Anxiety
▸ Depression
▸ Grief and loss of attachments
▸ Lack of social support
▸ Personality type
▸ Other stressors

POSSIBLE BIOLOGICAL MECHANISMS
▸ Stress neurotransmitters
▸ Clotting and thrombosis
▸ Damage to blood vessel walls
▸ Inherited conditions
▸ Inflammation
▸ Hormone imbalance

POSSIBLE BEHAVIORAL RISK FACTORS
▸ Lack of exercise
▸ Medication non-compliance
▸ Bad nutrition
▸ Obesity
▸ Poor sleep hygiene
▸ Smoking

POSSIBLE IMPACTS AS A RESULT
▸ Loss of confidence and sense of invincibility
▸ Reduced exercise stamina
▸ Loss of financial security
▸ Loss of independence
▸ Loss of sexual interest
▸ Illness
▸ Unemployment

DEPRESSION ⟷ **HEART DISEASE**

It is estimated that about 20 percent of healthy Americans and close to 50 percent of those being treated for heart disease are type D personalities. Studies suggest that the link between personality styles like A and D with an elevated risk of heart disease appears to be related to stress levels and inflammatory response.

THE UNHOLY ALLIANCE OF DEPRESSION AND CARDIAC DISEASE

Two stress-related chronic diseases—depression and heart disease—start with the metabolic syndrome. They lower the vulnerability threshold for each other and, thus, are common co-morbidities in the modern world. Indeed, the World Health Organization lists coronary heart disease and depression as two of the most disabling conditions found around the world.

Lifestyle

Cardiovascular disease is the major cause of what is called "the global burden of disease." In other words, it causes a greater degree of mortality, morbidity, and disability of any other disease. Thus, it is a public health challenge of the first order and all of us need to find ways to improve heart health in ourselves and in our communities. The risk factors for heart disease often stem from our lifestyle. These include an unhealthy diet leading to overweight, sedentary lifestyle, tobacco use, and alcohol and drug abuse. Poor sleep hygiene is also a menace to cardiac health. When psychosocial stress is added to these behavioral risk factors, it negatively affects the heart for the reasons discussed above.

The first inkling that your body is reacting poorly to these risk factors in the setting of stress is the emergence of the "metabolic syndrome." You may recall that the metabolic syndrome consists of the following problems: truncal obesity, high blood pressure, high levels of lipids in the blood, and the development of relatively insensitive receptors for insulin, making glucose metabolism a larger challenge for the body. The risk factors for metabolic syndrome include all the maladaptive behaviors mentioned above. They all conspire to produce what is called "oxidative stress" (a process we learned about in Chapter Two) and which is a common theme throughout this book.

Therefore, when we are stressed and add behavioral risk factors to the mix, our cells need to work overtime to manage more and more metabolic demands, leading to oxidative stress and diminished reserve capacity in the engine rooms of our cells—the mitochondria. The byproduct metabolites of oxidative metabolism then build up in the cell, leading to what is called free radical damage. This kick starts cellular pathways that produce an inflammatory response syndrome and cellular aging, resulting ultimately in accelerated cell death.

Cardiovascular disease

Cardiovascular disease results from a vulnerability you may have toward developing atherosclerosis—commonly called "hardening of the arteries"—which is due to the build-up of waxy plaque in the blood vessels. Your doctor may have checked your cholesterol levels from time to time. This is because a certain type of cholesterol, called low-density lipoprotein, can build up in the space between the cells that line your heart's blood vessels and the smooth muscle lining of the heart itself. Immune cells called macrophages, the white blood cells of your rapid-acting innate immunity, in the throes of an inflammatory response, seek out the low-density lipoprotein, gobble it up, and proceed to become foamy cells. These cells eventually burst like soap bubbles releasing a host of toxic substances including the oxidative metabolites mentioned earlier.

Once this process gets going, it causes tearing in the coronary artery's cell wall. Platelets will respond in an effort to seal the breach. The ultimate result if this problem becomes chronic will be the development of a large plaque of coagulated platelets called a "thrombus." This spells trouble, since this thrombus blocks oxygenated blood from reaching the heart muscle, leading to muscle damage due to lack of oxygen, in a process known as "ischemia." If you suffer significant muscle cell death in this way, you will have what is called a myocardial infarction, commonly known as a heart attack. A myocardial infarction can often damage your heart, leading to reductions in the cardiac output of blood to the tissues and even heart failure. A myocardial infarction will cause scarring in the heart when it heals, and this scarring will make you more vulnerable to electrophysiological accidents (more common under stress response conditions), such as ventricular arrhythmias that may lead to your dropping dead from a cardiac arrest, in what is referred to as sudden cardiac death. The heart ventricle on the left is the large chamber that pushes oxygenated blood around the body, so when it is not in synchrony because of an arrhythmia, the output of the heart is compromised and tissues, especially the brain and kidneys, can suffer greatly.

Healthy changes

Lifestyle change is therefore important in order to avoid coronary heart disease. Healthy changes can reduce stress and improve our capacity to stay well. Doing this will divert cells away from mounting a stress-induced inflammatory response and promoting aging pathways. By adopting healthy lifestyle changes in the following areas—exercise, diet, sleep hygiene, stress reduction, and resilience enhancement—we can lower blood pressure, heart rate and oxygen consumption, and reduce body mass index. This reduces the strain placed on the cell to process glucose and oxygen, and this improves cellular health and advances our overall performance in the long run.

Cells both in the brain and in the heart stand to benefit, if these healthy lifestyle behaviors are adopted. In some cases, doing so has even been shown to reverse ischemic heart disease.

Stress reduction is an integral part of lifestyle management. Practicing mind–body approaches such as meditation can alter the stress-related nervous system imbalance between the sympathetic and parasympathetic systems. Integrating these practices into our lives can provide us with better blood pressure control, enhancement in our insulin receptor sensitivity, reduction of lipid peroxidation, and delay in cellular aging as researchers at the Benson-Henry Institute for Mind Body Medicine and elsewhere have shown. The stress-related metabolic syndrome can thus be improved, and along with this, our vulnerability to cardiac disease and heart attacks will be diminished.

HARDENING OF THE ARTERIES

Lumen (vessel interior)

Artery wall

Fatty plaque

ATHEROSCLEROSIS
A colored scanning electron micrograph (SEM) of a cross-section through a human coronary artery of the heart showing atherosclerosis. This is a build-up of fatty plaques on the walls of arteries. The artery wall is red; hyperplastic cells are pink; fatty plaque is yellow. A large section of the artery width is blocked as only a tiny portion of the lumen (the vessel interior) is seen (blue, far right). The blockage is partly plaque, but mostly due to the hyperplastic reaction of cells in the artery wall that have multiplied. Atherosclerosis leads to irregular bloodflow and blood clots, which can block the coronary artery resulting in a heart attack.

BODY FAT DYNAMICS
Body fat and the stress response

If we develop the features of metabolic syndrome through the effects of stress on our genes, they will conspire together to compound our disease risk. For example, it now appears that obesity itself is associated with a chronic, innate inflammatory response that causes low-grade immune activation in fat tissue. This persistent immune activation sets us up for a variety of diseases including heart disease.

Body Mass Index (BMI)
Many people all around the world struggle with their weight. The terms "overweight" and "obesity" do not mean the same thing. These terms represent different points on the weight continuum, ranging from underweight to obese. Where someone falls on this path is determined by a measurement called the Body Mass Index.

BMI is a measure that relates weight to height. BMI is usually considered an effective way to determine whether a person is overweight or obese. According to the National Heart, Lung, and Blood Institute (NHLBI) in the US, a BMI from 18.5 to 24.9 is considered normal while a BMI of more than 25 is considered overweight though muscular, weighty people among us may have BMIs this high and not be overweight. A BMI above 30 usually reflects obesity, and a BMI greater than 40 is usually considered to be in the morbidly obese range.

Obesity stimulates an inflammatory state, a core feature in many stress-related chronic non-communicable diseases, including type 2 diabetes, hypertension, and heart disease. Fat (adipose) tissue itself seems to play a key role in the pathogenesis of these disorders. And macrophages, those white blood cells that set the innate immune response in motion, play a key role in engineering the inflammatory response syndrome in adipose tissue.

White adipose tissue
How does this occur? We all have white adipose tissue (WAT) in our fat stores and it is of particular importance as the storage site for lipid-based energy. We benefit from having this tissue as an energy storage depot to draw on when food is scarce. In obesity, the white adipose tissue increases production and secretion of protein molecules called cytokines that promote inflammation. These molecules not only have local effects on white adipose tissue physiology, but also produce far-reaching systemic effects on other organs when they circulate in the blood stream. Furthermore, macrophages from the bloodstream infiltrate white adipose tissue in obese individuals and become the major source of locally produced pro-inflammatory cytokines that keep kindling the fire of a harmful, persistent immune-activated state. As a result, inflammation is chronically present in the fat as well as the vasculature of many patients with atherosclerotic heart disease and type 2 diabetes.

Stress hormones
Chronically stressful states may be a major culprit in this overall inflammatory process. The mediators of this effect are the stress hormones, namely norepinephrine (noradrenaline), epinephrine (adrenaline), and cortisol. Together with components of other substances produced as a result of the breakdown of fat, such as cytokines and free fatty acids, these transmitters will activate a special chemical facilitator of gene activation called a transcription factor. The specific transcription factor that is stimulated in an obese individual is called Nuclear Factor kB (NF-kB) and it is turned on in macrophages, fat, and endothelial cells. Stress also activates NF-kB in the brain, particularly in the prefrontal cortex. NF-kB is a key link between stress and disease.

NF-kB encourages the production of certain receptors on these cell types, which, when engaged, produce a cascade of inflammatory reactions resulting in the innate cell-mediated immune system response. While this ancient immune response is extraordinarily important in protecting us against microbial and traumatic threats, it has in the course of evolution also become subject to psychosocial stress. And since psychosocial stress tends to be ubiquitous in our modern society, stress-induced immune activation may be constantly left on. This will sap us of our energy and strength and lead to a variety of illnesses.

The inflammatory process is most pronounced in the white adipose tissue as well as blood vessels, and is involved in stress-induced metabolic events, culminating in the metabolic syndrome, the effects of which precede and comprise the major risk factors for heart disease, type 2 diabetes, and other disorders.

In a way therefore, obesity, which is reaching epidemic proportions not only in high-income countries but also in low- and middle-income nations, is both the chicken and the egg of the inflammatory response, which is so common as the precursor for chronic, non-communicable illnesses such as cardiac disease.

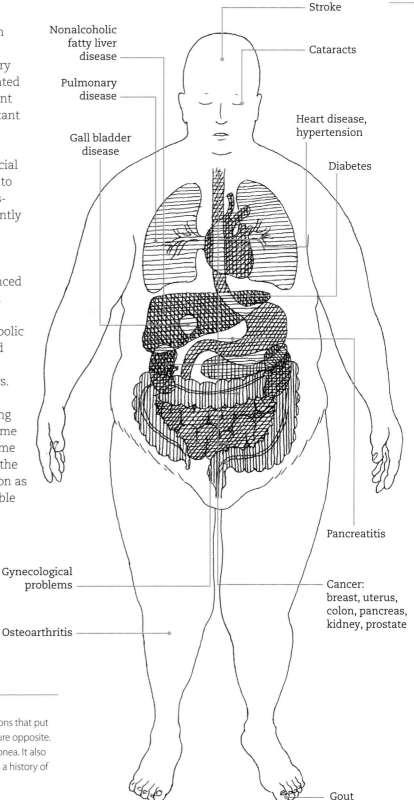

MEDICAL COMPLICATIONS OF OBESITY
Obesity can over time lead to devastating medical conditions that put stress on the body—many of which are shown in the picture opposite. In addition, obesity is a major cause of obstructive sleep apnea. It also correlates with a higher incidence of depression, as well as a history of toxic stress in childhood.

THE IMPORTANCE OF INSULIN EFFECTIVENESS
Stress and blood sugar levels

The road from chronic stress to obesity and to insulin resistance is a downhill slippery slope. Insulin is a hormone produced in your pancreas. It enables glucose to be processed from your bloodstream. Whereas type 1 diabetes stems from a lack of insulin production, type 2 develops from poor insulin receptor functioning.

Resistance to insulin at the receptor site leads to impaired glucose tolerance and the elevations in glucose levels in the blood that result will predispose to type 2 diabetes. And, of course, it is well known that type 2 diabetes can present simultaneously with other problems such as obesity, cardiovascular disease, and cerebrovascular disease.

It is fascinating to consider the benefits of weight loss in this regard. It is known to improve all dimensions of the metabolic syndrome, so it helps to lessen an individual's risk of heart disease, stroke and diabetes. This benefit may follow from the fact that weight loss is associated with decreased infiltration of white adipose tissue by macrophages with the subsequent decreased activation of NF-kB pathways. This situation will decrease that inflammatory response cascade described above. Since NF-kB is the transcription factor at the gene level responsible for activating the genes that produce the protein cytokines necessary for the inflammatory response, reducing NF-kB activation has the potential to be health promoting at the source.

INFLAMMATORY RESPONSES IN THE BODY

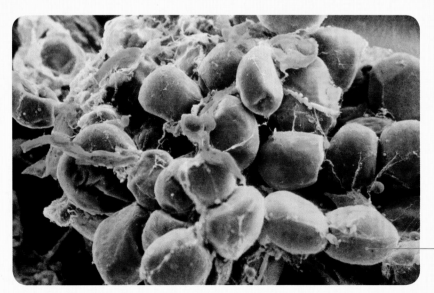

FAT CELLS
Colored scanning electron micrograph of fat-storing cells (yellow), also known as adipocytes, which build up the adipose connective tissue. Fat not used in metabolic processes is channeled toward these cells through small capillaries; a few of them are seen as blue tubules. The body's hormonal reaction to chronic stress, as well as the breakdown of fat itself, activates NF-kB in these adipocytes. Macrophages, which infiltrate adipose tissue, also produce NF-kB. The result of NF-kB production is an inflammatory response that can lead to metabolic syndrome..

Fat cells

Fat cells

The pathogenesis of insulin resistance is complex and is the result of multiple factors. Fat cells as well as infiltrating macrophages provide a potpourri of signaling molecules that impact the process. These cells, in the setting of obesity, produce excessive amounts of weight affecting chemical messengers like the hormones leptin and resistin in addition to the inflammatory cytokines already mentioned. The hormone resistin can promote insulin resistance, while leptin is thought to regulate macrophage activation and cytokine production.

Adiponectin

Another chemical messenger to be considered is the peptide adiponectin. Adiponectin levels are lower in patients with obesity-related insulin resistance, type 2 diabetes, and heart disease. It is now thought that adiponectin counteracts harmful pro-inflammatory effects on arteries and probably protects against the development of arteriosclerosis.

Obesity, therefore, may be reconceptualized in terms of an inflammatory response. It is this immune activation that predisposes to insulin resistance.

These molecular interactions alter insulin receptor sensitivity by mediating the insulin-signaling pathway often in conjunction with modulation of the inflammatory response. This is dangerous, because when our insulin receptor loses affinity for the insulin hormone, the stage is set for an eventual slide into type 2 diabetes and all its attendant illnesses.

We now know that meditative and other mind–body approaches will also change our inflammatory profile. Research in the US from the Benson-Henry Institute at Massachusetts General Hospital, UCLA, and the University of Miami, has shown that by practicing a variety of mind–body approaches that reduce stress, we can activate certain genes and deactivate others in the white blood cell precursors of macrophages. In the process, our health-promoting cellular profile becomes favored and this means that our cell-mediated immune activation will be dampened.

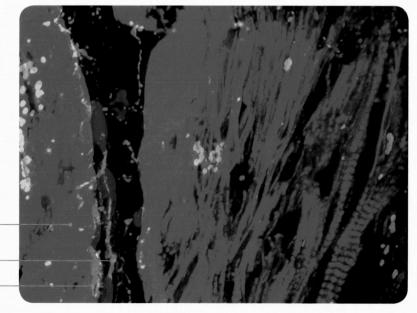

NF-kB PATHWAY
Fluorescence micrograph showing the transcription factor NF-kB (in red) in heart muscle tissue from a patient with heart failure. The action of the muscle's contractile fibers is shown in green and the blue areas are cell nuclei. NF-kB is a protein complex that activates genes associated with cell death (apoptosis) processes and immune system regulatory pathways.

Muscle's contractile fibers ———

Transcription factor NF-kB ———

Cell nuclei ———

HEART DISEASE IS IMMUNE DISEASE

Heart disease is a stress-related disease

We have seen that psychosocial stress contributes to the development of cardiovascular disease as well as to many other illnesses. Today, most people are aware that stress can be dangerous to heart health.

It is now increasingly clear that psychosocial stress, especially that associated with negative thinking and depression, can be *translated* through the effects of stress mediators, into metabolic overactivation and then *processed* at the cellular level as oxidative stress. Oxidative stress is a term used to describe the cellular consequences of the build-up of toxic substances, stemming from the need for your cells' mitochondria to over-process metabolites (oxygen and glucose) when stressed physically or psychosocially. Physical and mental stress creates a challenge to an organism's stability (homeostasis), triggering a flexible physiological response (allostasis) to the changing circumstances. It takes metabolic energy for the brain to counter stressors, while also returning to a normal physiological state when the threat is eliminated. This can lead to the metabolic wear and tear at the cellular level, which increases our vulnerability to disease when stress is unremitting, overwhelming, or merely perceived by us as such.

It is our brain's stress response systems, consisting of the amygdala-driven sympathetic nervous system, and the HPA cortical axis, that serve as conduits to the body's end organs. These arms of the stress response system respond to what our brain processes as a challenge or a threat. We now know that psychosocial stress alone will ignite brain-derived inflammation without an infectious or physical trauma stimulus. This state stimulates target tissues like the heart to alter their metabolisms to match this new state of affairs, increasing the cells' allostatic loads. If the psychosocial stress is chronic, cellular oxidative stress can turn this process into a disease-producing one. The fact that C-reactive protein, a marker of the inflammatory response, is now considered an independent risk factor for heart disease supports the notion that inflammation plays a role in its genesis.

The pro-inflammatory transcription factor NF-kB is a potential critical bridge between stress and oxidative cellular activation and plays a pivotal role in vascular disease. Neurotransmitters evoked by psychosocial stress stimulate activation of NF-kB. The resulting inflammation may directly target the functioning of our blood vessel lining in the coronary arteries and thus represent an additional risk factor for cardiovascular disease.

Allostatic loading itself can lead to oxidative stress with the resultant production of free radicals and stress-sensitive heat shock proteins. The heat shock proteins were discovered when high temperatures physically stressed cells; however, they surface under other stress conditions as well. They serve as molecular connectors that enhance macrophage activation and set the inflammatory response in motion, leading to production of pro-inflammatory and cell-damaging mediators.

When stress is removed, NF-kB usually returns to its baseline level of activation within sixty minutes or so. However, some individuals are slower to down regulate, prolonging NF-kB activation. This differential response to stress stimulation suggests variability in stress perception, and may be a product of gene activation differences.

EARLY INFLAMMATION

A pocket of fatty, soft plaque begins to build up

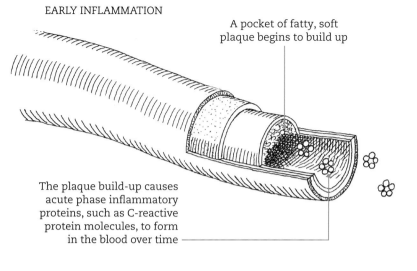

The plaque build-up causes acute phase inflammatory proteins, such as C-reactive protein molecules, to form in the blood over time

ADVANCED INFLAMMATION

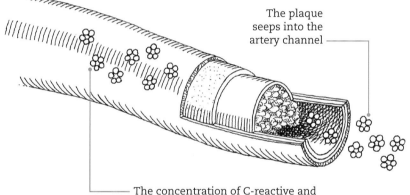

The plaque seeps into the artery channel

The concentration of C-reactive and other inflammatory proteins rises with the severity of white blood cell-mediated inflammation

HEART DISEASE AND CHRONIC INFLAMMATION
Macrophages are the maestros of the cell-mediated innate immune response. They are key in the development of cardiovascular disease. Platelets complicate the picture by producing a thrombus with plaque build-up that kindles more inflammation and eventually limits blood flow.

We can learn from a study that looked at 2,320 men who had suffered a heart attack. When these men had high life stress, or were socially isolated, they were significantly more likely to die within three years of a heart attack. Furthermore, when they had both high stress and poor social support, they had a very high mortality rate.

Cardiac patients like those described often have higher levels of depression and anxiety as "allostatic load disorders," and are more susceptible to psychosocial stress, and to further cardiac events. These individuals may benefit most from finding ways to relax as a counterbalance to stress, which may enhance their resilience to disease. When we can do this, we can diminish activation of the gene sets responsible for accentuating disease-producing immune activation.

For all the reasons mentioned in this chapter, heart disease represents perhaps the best example of a stress-related illness besides those associated with the brain. The brain and the heart must be constantly attuned to one another if a harmonious life is to be attained and maintained, and if the risk of heart disease is to be controlled.

Chapter Four

STRESS AND THE IMMUNE SYSTEM

Stress can be extremely detrimental to our health and well-being in numerous ways. Many harmful effects of stress are produced through its effects on the immune system, the complex biological network of cells, tissues, and organs within our bodies that defends against external infectious agents, and from the growth of tumor cells. To protect the survival of the human body, the immune system must detect various intruders such as bacteria and viruses, and distinguish them from our bodies' own healthy tissues. As many of the intruders are able to evolve over time to avoid detection, the immune system adapts continually to be able to recognize and fight them. Thus, the immune system is vital for the survival and well-being of our bodies. In this chapter, we will discuss various mechanisms through which stress is related to aging, poor memory, and various physical diseases and psychological illnesses.

YOUR IMMUNE DEFENSES
The shotgun and the sniper

The immune system protects the human body with multiple layers of defense. The first layer consists of physical barriers, preventing pathogens from entering the body. If a pathogen passes the physical barriers, it will encounter the innate immune system, which provides an immediate but non-specific response. If the innate response fails to stop the pathogen, another layer of protection, the adaptive immune system, will be activated by the innate response. The adaptive immune system has immunological cells to preserve memory after their initial response to a specific pathogen, leading to a faster and stronger response to subsequent encounters with that same pathogen. Using the analogies of weapons, the innate immune system is like using a shotgun to attack all pathogens, and the adaptive immune system is a sniper approach to target specific pathogens.

Innate immune system

The innate immune system is found in nearly all forms of life. Even simple unicellular organisms possess a rudimentary immune system, in the form of enzymes that protect against infections. The innate system is the dominant system of host defense in most organisms, but it does not have immunological memory, so it cannot provide long-lasting immunity against specific pathogens. It offers an extensive, but less efficient way of host defense to the many microorganisms that can cause disease.

There are several surface barriers to protect our bodies from infection, including mechanical, chemical, and biological barriers. Mechanical barriers, such as skin, are the first line of defense against infection. Chemical barriers such as the enzymes in saliva, tears, and breast milk are antibacterial. A good example of a biological barrier is the bacteria found in our genitourinary and gastrointestinal tracts. They directly compete with pathogens for food and space,

decreasing the chance that pathogens will reach levels sufficient to cause disease. Research has shown that reintroduction of bacteria, such as the lactobacilli normally found in unpasteurized yogurt, helps restore a healthy balance of bacteria populations in the intestines of children.

When pathogens successfully pass the surface barriers and enter an organism, they will trigger the pattern recognition receptors present in many of the immune cells including macrophages, dendritic cells, and mastocytes. Once activated, these cells release inflammatory mediators, including cytokines, which trigger the non-specific immune response. Common cytokines include interleukins, which are responsible for communication between white blood cells; chemokines, which promote movement of immune cells; and interferons, which have anti-viral effects. These cytokines and other chemicals recruit immune cells to the site of infection to facilitate the removal of the pathogens and promote healing of any damaged tissue. These activities are responsible for the clinical signs of inflammation, which include redness, swelling, heat, and pain caused by increased blood flow into tissue.

ORGANS OF THE IMMUNE SYSTEM
Many organs play important roles in the immune system. Bone marrow produces immune cells, some of which differentiate into T lymphocytes in the thymus. The spleen stores immune cells, while tonsils and adenoids contain lymphocytes that activate the immune system when they come into contact with pathogens. The lymph nodes filter and clean the lymph, which flows into lymphatic vessels which drain into the blood stream. Peyer's patches in the small intestine contain mainly B cells, which are a type of lymphocyte involved in antibody production (see also pages 72–3). The appendix also has gut-associated lymphoid tissues, although their function is less clear.

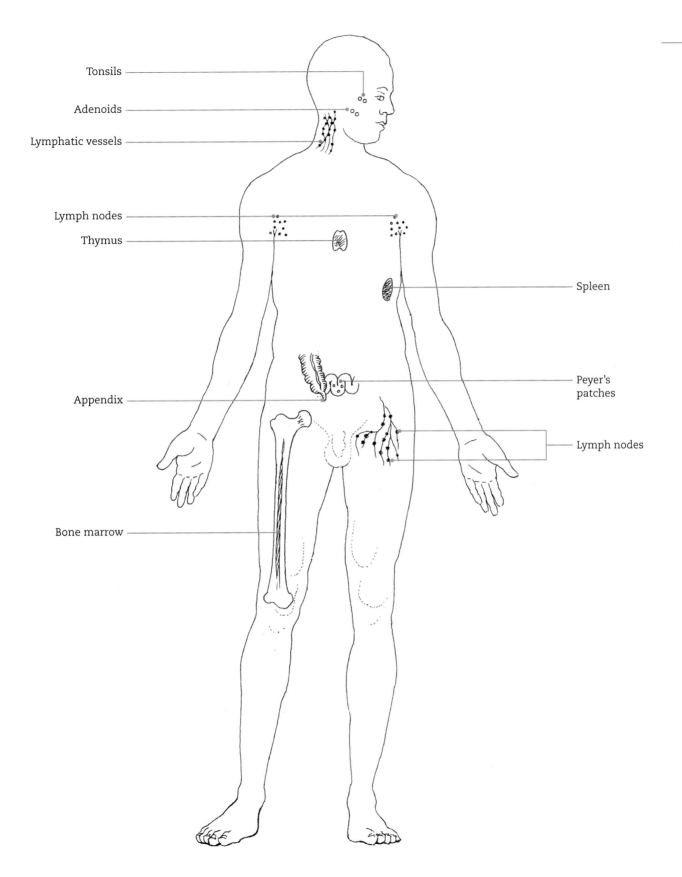

Tonsils

Adenoids

Lymphatic vessels

Lymph nodes

Thymus

Spleen

Appendix

Peyer's patches

Lymph nodes

Bone marrow

The complement system

Another major component of the innate immune response in the blood is the complement system, which is a part of the immune system that helps or complements the ability of antibodies and various white blood cells to clear pathogens from an organism. The complement system consists of a number of small proteins found in the blood that normally circulate as inactive precursors. When complement proteins bind to a pathogen, they activate a series of enzyme activities, which creates a catalytic cascade that attacks the surface of the pathogen and eventually kills it.

Adaptive immune system

The adaptive (or acquired) immune system provides a stronger immune response compared to the innate immune system. Because of its immunological memory, which allows specific antigens (foreign substances that produce antibodies) to remember each pathogen it encounters, the adaptive immune system can generate tailored responses to eliminate specific pathogens or pathogen-infected cells. Memory cells maintain these tailored responses. If a pathogen infects the organism more than once, the specific memory cells will respond in a focused and efficient way to eliminate the pathogen. However, there is lag time between the exposure of the pathogen and the maximal response. This is because the antigen-specific response needs to go through a process called antigen presentation for the initial recognition of the pathogen. As a result, the organism may show signs of harmful effects by the pathogen during the lag period before the adaptive immune response is mobilized to eliminate the pathogen.

Similar to the innate immune system, the adaptive immune system also has white blood cells, which are called lymphocytes. The two major types of lymphocytes are B cells that are involved in antibody production, and T cells that are involved in cell-mediated immune response. When B cells and T cells

THE CELL'S IMMUNE RESPONSE

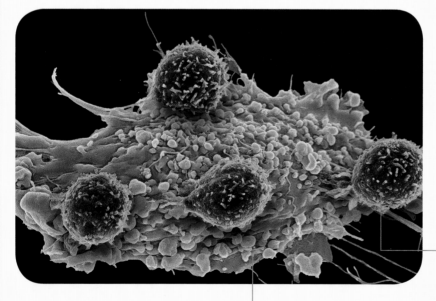

T CELL

This colored scanning electron micrograph (SEM) shows T cells (red), or T lymphocyte cells, attacking a cancerous cell (green). T cells recognize a specific site on the surface of pathogens or foreign objects, bind to it, and produce antibodies or cells to eliminate that pathogen or foreign object. As we shall see, persistently raised cortisol levels diminish the function of T cells and other agents of adaptive immunity.

T-lymphocyte cell

Dendritic processes of the pathogen

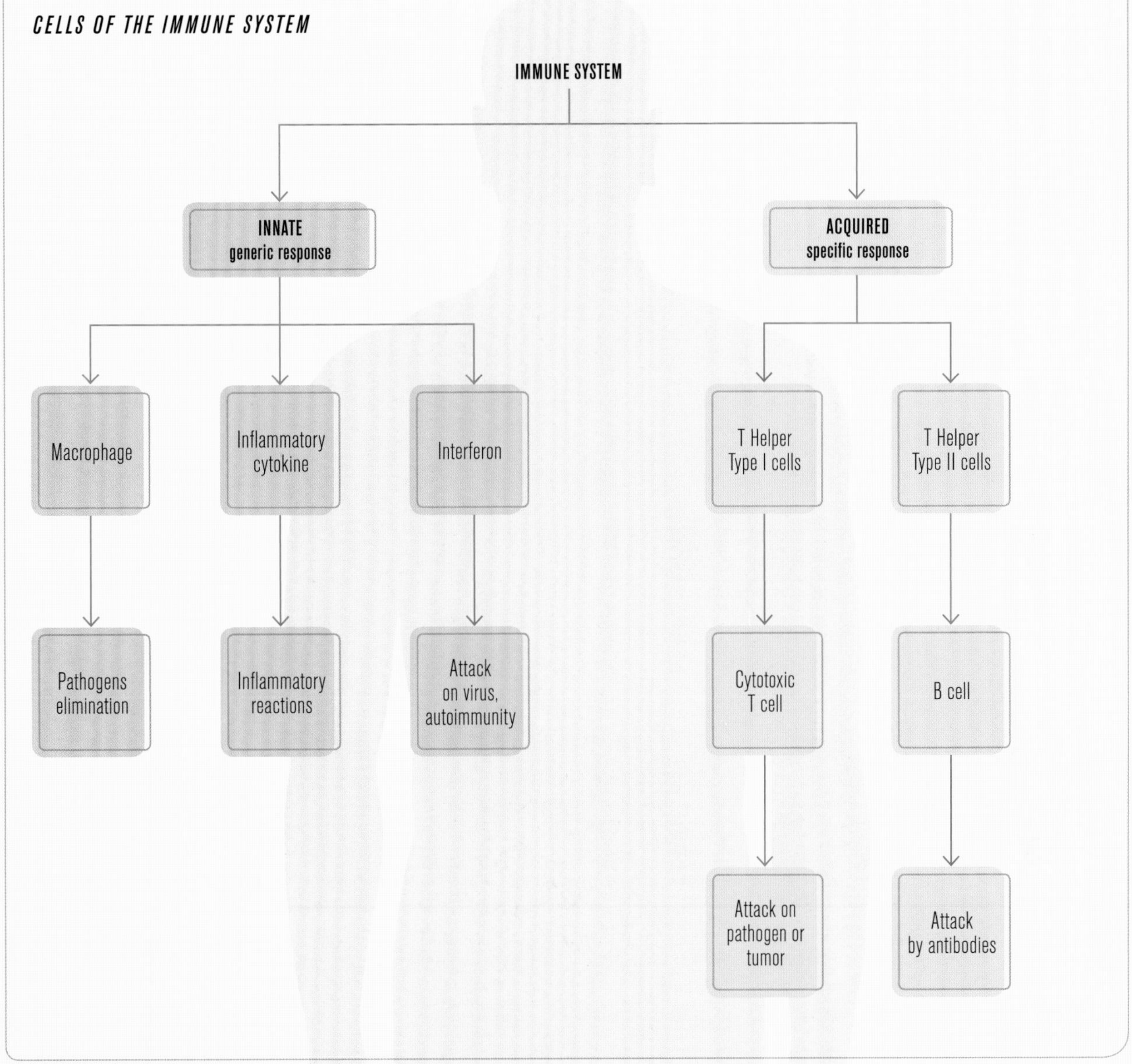

INNATE AND ACQUIRED IMMUNE SYSTEM
This chart shows how the activation processes of the innate immune system along with the acquired immune system help to protect against bacterial infection.

replicate, some of their offspring become memory cells, which can remember specific pathogens encountered and allow the adaptive immune system to give a strong response when a pathogen is detected again. This process is known as immunological memory and it can be either a passive short-term memory or an active long-term memory. The latter is acquired after the infection activates B and T cells. It can also be created artificially through vaccination.

THE IMMUNE SYSTEM RESPONSE MOBILIZERS

The immune system's response to stress

Research in the past thirty years has established that stress prompts the brain to activate both the sympathetic nervous system and the hormonal system that includes the hypothalamus-pituitary-adrenal (HPA) axis and other endocrine organs, such as the thyroid, and testis or ovaries.

When the brain perceives environmental and psychological stressors, this information is first identified through the hippocampus and amygdala, which then stimulate the hypothalamus to activate the sympathetic nervous system. This nervous system stimulates the adrenal glands to secrete epinephrine or norepinephrine. These hormones cause pupil dilation, and increases in the rate and force of contraction of the heart, blood sugar levels, and blood supply to skeletal muscles. These changes engage our fight-or-flight response and prepare the body for violent muscular action by moving blood away from areas that are not involved with physical movements, such as the digestive organs.

There is also a corresponding hormonal response to stress through the HPA axis. The hypothalamus secretes the corticotropin-releasing hormone, which induces the pituitary gland to secrete adrenocorticotropic hormone (ACTH). This hormone stimulates the secretion of cortisol from the adrenal gland. Cortisol affects virtually the entire body, including the immune response, memory, and the transformation of fatty acids into available energy, which prepares muscles throughout the body for response. Cortisol weakens the immune response by slowing the production of T cells, and by preventing some T cells from functioning properly. It does this by inhibiting the production of histamine, a chemical involved in the response to invasion by pathogens.

Chronic stress

People suffering from chronic stress are highly vulnerable to infection, because their immune response has been diminished. High levels of cortisol have been shown to negatively affect memory through its action on the hippocampus, the region of the brain where memories are processed and stored. Excess cortisol overwhelms the hippocampus and actually causes atrophy. The HPA axis is the primary stress management system in the human body, and it enables us to maintain homeostasis in response to both physical and mental challenges by controlling the body's cortisol levels.

The impact of stress on immune system functioning depends on the severity and duration of the stressful event. Herbert and Cohen reviewed thirty-eight studies of stressful events and immune function among healthy adults and found that stress increases the number of white blood cells and decreases the function of B and T cells. Studies on acute sleep loss showed that it confuses the HPA axis, and results in elevated plasma cortisol levels by up to 45 percent after sleep deprivation, an increase that has implications including compromised immune function and cognitive impairment.

On the other hand, other research studies on stressors and human immunity showed that acute time stress (for example, public speaking) can boost the body's natural immunity by increasing the number of natural killer cells and large granular lymphocytes, as well as the production of pro-inflammatory cytokines. Short-term naturalistic stressors (for example, exams making students feel stressed) decrease cellular immunity. When stress becomes chronic, the antibody-mediated immunity

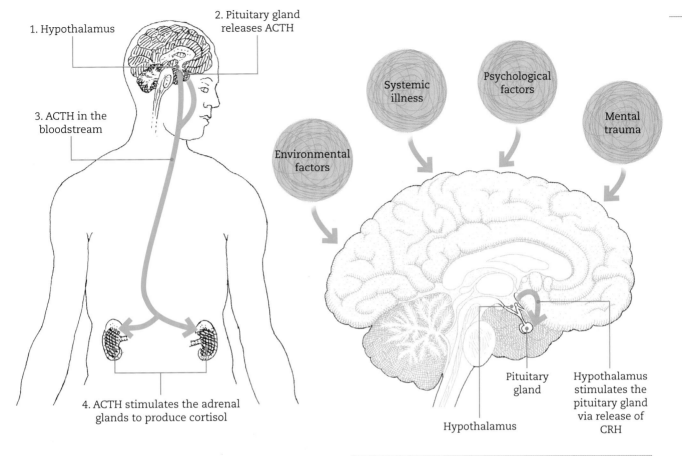

1. Hypothalamus

2. Pituitary gland releases ACTH

3. ACTH in the bloodstream

4. ACTH stimulates the adrenal glands to produce cortisol

Environmental factors

Systemic illness

Psychological factors

Mental trauma

Pituitary gland

Hypothalamus

Hypothalamus stimulates the pituitary gland via release of CRH

STRESS AND THE HPA AXIS

The HPA axis begins with the hypothalamus releasing corticotropin-releasing hormone (CRH). This causes the pituitary gland to release adrenocorticotropic hormone (ACTH) into the blood, which causes the adrenal glands to release cortisol.

may be dampened while innate immunity is activated. Chronic innate immunoactivation can bring harmful effects to the body. In addition, the muscles in the body are put in a constant state of readiness. When muscles are tight and tense for long periods of time, this may trigger other reactions in the body and promote stress-related disorders. For example, both tension-type headaches and migraine headaches are associated with chronic muscle tension in the area of the shoulders, neck, and head.

High levels of stress, even over relatively short periods of time and in vastly different contexts, tend to produce similar results: prolonged healing times, a reduction in the ability to cope with vaccinations, and heightened vulnerability to viral infection. The long-term, constant cortisol exposure that is associated with chronic stress produces further symptoms, including impaired cognition, decreased thyroid function, and accumulation of abdominal fat, which may have implications for cardiovascular health.

How to counter stress

The brain serves as our allostasis monitor to maintain stable physiologies in the face of changing external and internal environmental challenges. An effective way to counter the effects of stress is relaxation. Relaxation has been shown to increase the secretion of beta-endorphin from the brain, and reduce chronic immunoactivation by decreasing NF-kB pathways as described in Chapter Three. Relaxation techniques have been shown to effectively reduce muscle tension, decrease the incidence of certain stress-related disorders (such as headaches), and increase a sense of well-being. Relaxation has also been shown to reduce recurrence and mortality rates of cancer patients, and prolong their survival time.

STRESS AND GROWING OLD
How stress affects the aging process

Aging is part of life that everyone has to experience. Even with help from the most advanced scientific technologies, human beings are unable to reverse the process of getting old. There are certain cognitive functions that are particularly vulnerable to age-related deficits, such as the decline in cognitive control, reduced processing speed, as well as impaired selective attention. Generally, age-related cognitive deficits are observed in tasks that are difficult and/or require shifting from one task to another. Individuals who show memory problems that exceed what is normally associated with aging are diagnosed with mild cognitive impairment. They are more forgetful, feel increasingly overwhelmed by making decisions and planning, and lose their train of thought, and show increasingly poor judgment. Some of these individuals with mild cognitive impairment may proceed to more severe forms of dementia such as Alzheimer's disease and have interferences with their day-to-day life and usual activities. Evidence suggests aging, stress, and the immune system interact with each other, and that this interaction influences the cognitive functioning of elderly individuals.

Research has shown that normal aging is associated with the development of low-grade chronic inflammation in the central nervous system (CNS), and this chronic immune activation can disrupt cognitive processes and even be toxic to brain cells. The brain's resident immune cells, microglia, are the primary mediators of the innate immune response in the brain, and play a crucial role in the age-related increase of inflammation in brain cells. In a normal adult brain, microglial cells remain in a resting surveillance state, which means the cellular processes constantly survey for any indication of injury or infection. Once a threat is found, the microglia are activated and quickly respond to combat the threat.

A number of studies observing both humans and animals have confirmed that aging enhances pro-inflammatory cytokines and that aging increases microglia cell proliferation and induces morphological changes, both of which are important features of microglia activation. Age-related microglia activation is likely to contribute to cognitive deficits in the aging process. Post-mortem analysis of cytokine level in the brains of Alzheimer's patients showed a marked increase in pro-inflammatory cytokines, suggesting a potential link between inflammation and cognitive decline.

Theories of aging

There are a number of theories and explanations for the basis of aging. The American biogerontologist Denham Harman (1916–2014) proposed the "free-radical theory of aging" in the mid-1950s. He suggested that oxidative enzymes catalyze the molecular oxygen in cells and then produce free radicals. During respiration, these free radicals induce deleterious effects on cell components and connective tissues, causing cumulative damage over time, which ultimately results in aging and death.

Later, in the 1970s, this theory was expanded to involve mitochondria, the structure in the body's cells which turns food into usable forms of energy. The mitochondrial free radical theory of aging suggests that mitochondria are both producers and targets of reactive oxidative species. In this theory, oxidative stress attacks and impairs mitochondrial function, leading to aging. Activated microglia are the most abundant source of free radicals in the brain. Microglia-derived radicals produced as a result of increased inflammation during aging may be a cause of oxidative damage and brain cell death.

Another theory on aging examines the relationship between aging and the tiny cell clocks called telomeres that determine how long cells will live. Telomeres are little caps at the end of chromosomes that prevent loss or injury to genetic information during cell division. Cells with long telomeres live longer. Short telomeres are linked to a wide range of human diseases, such as coronary heart disease, osteoporosis, and HIV infection. Researchers found that people subjected to chronic stress tend to have shorter telomeres. Another study showed that cortisol suppresses telomerase (a compound that protects telomeres), leading to early cell aging and distorted cell replicas, which could lead to cancer and other severe diseases.

Dementia

The effects of prolonged elevations of cortisol as part of the HPA axis response to chronic stress may be a cause of cognitive decline in the elderly. This response targets specific brain regions such as the hippocampus, amygdala, and prefrontal cortex. The hippocampus, which receives the most influences from the HPA axis response, is critical for certain types of memory and is the initial site of pathological changes in Alzheimer's disease. Research suggests that older adults who respond to chronic stress with elevated cortisol levels are at greater risk for developing dementia.

EFFECTS OF DEMENTIA ON THE BRAIN

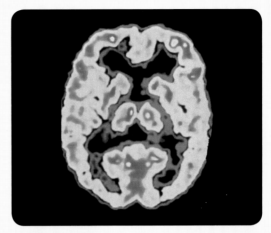

High brain activity

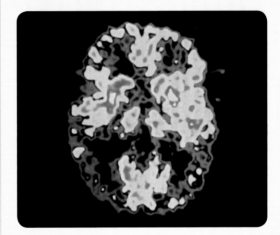

Low brain activity

ELEVATED STRESS MAY LINK TO THE DEVELOPMENT OF DEMENTIA

The two images above are colored positron emission tomography (PET) scans of the brain. The upper image is of a normal person, the lower image is of a patient with dementia who is suffering from Alzheimer's disease. The upper image of the normal patient shows more high brain activity areas (red and yellow); and the lower image of the patient with dementia shows more low activity areas (blue and black).

STRESS, DEPRESSION, AND OTHER PSYCHIATRIC DISORDERS

Stress and its effect on your mental heath

There is a lot of evidence which suggests that stress and depression are interrelated. It has been proposed that many patients of depression are in a state of chronic inflammation, with prolonged and excessive mobilization of white cells and chemicals (for example, C-reactive proteins and cytokines) involved in inflammation. As a result of chronic inflammation, patients suffer from fatigue, loss of appetite, and difficulty concentrating, sickness symptoms typically seen in patients with depression. The degree of elevation of certain pro-inflammatory cytokines has been shown to correlate with the severity of depressive illness.

There is a growing body of evidence to suggest that commonly used antidepressant drugs also show anti-inflammatory properties. Pre-clinical studies have shown that common antidepressants possess anti-inflammatory actions. Studies have shown that antidepressants are able to suppress the production of pro-inflammatory cytokines. This finding strongly suggests that antidepressants can act to normalize the mechanisms responsible for increased cytokine production. Non-pharmaceutical treatments for depression also exert effects on the immune system. Exercise has been shown to have an anti-cytokine effect on the body, omega-3 fatty acids can lower levels of pro-inflammatory cytokines and raise levels of anti-inflammatory cytokines. St. John's Wort (*Hypericum perforatum*) is a flowering plant which also shares antidepressant and anti-inflammatory properties.

It is unclear how cytokines exert their effects on the brain, as they are large molecules that are generally unable to penetrate the blood–brain barrier, the "wall" that is formed by blood vessels separating what is in the blood stream and the brain cells. It has been postulated that cytokines may act at sites where the blood–brain barrier is weak or non-existent, or they could be transported actively into the brain. It is also possible that cytokines in peripheral circulation trigger the production or release of cytokines inside the brain. Further research is needed to provide a more definite answer to this question.

Stress and other psychiatric disorders

Post-traumatic stress disorder (PTSD) represents a condition that may arise shortly or many years after exposure to a serious traumatic event or injury. It is characterized by mentally re-experiencing the initial trauma and is frequently accompanied by symptoms such as anxiety, insomnia, nightmares, memory loss, behavioral changes; and increased susceptibility to infections, immune suppression, depression and—potentially—violent acts. As a result, patients with PTSD are chronically under stress.

Stress induces hormones including glucocorticoids, epinephrine, and norepinephrine that can have detrimental effects on neurons in the hippocampus, amygdala, and other parts of the brain. In addition, these hormones modulate the immune system via cytokine secretion, and result in cognitive and behavioral changes in patients with PTSD.

Delirium

Delirium is a common and severe neuropsychiatric syndrome; evidence suggests that one in eight hospital patients suffer from this condition. Symptoms of delirium include impairment in attention and cognitive abilities, reduced alertness and altered sleep-wake cycle, and psychosis. There are a lot of similarities between delirium and immunological and behavioral responses to infection such as fatigue, loss of appetite, sleepiness, and decreased energy. Some studies indicate that delirium can be regarded as manifestations of sickness behaviors due to systemic inflammatory immune responses in the body. The HPA axis is an important pathway that is activated in delirium states. In healthy individuals, inflammation activates the HPA axis, which then mobilizes energy resources to limit the inflammatory response.

However, age-related and stress-induced dysregulation of the HPA axis can make it more responsive to acute stress, which can result in high levels of cortisol persisting for weeks, which has been associated with delirium in elderly patients.

THE SYNTHESIS OF NOREPINEPHRINE IN A NERVE CELL
The neurotransmitter norepinephrine is made from another neurotransmitter, dopamine, in tiny bubbles called vesicles inside neurons. Dopamine is made from the amino acid tyrosine, which is a product of the breakdown of proteins from digested food. The norepinephrine is released across synapses (gaps) between neurons. In the receiving postsynaptic neuron, it effects changes in activation or behavior; in the amygdala, for example, it enhances the formation of long-term memories, and in the hypothalamus it is involved in originating the stress response via the HPA axis.

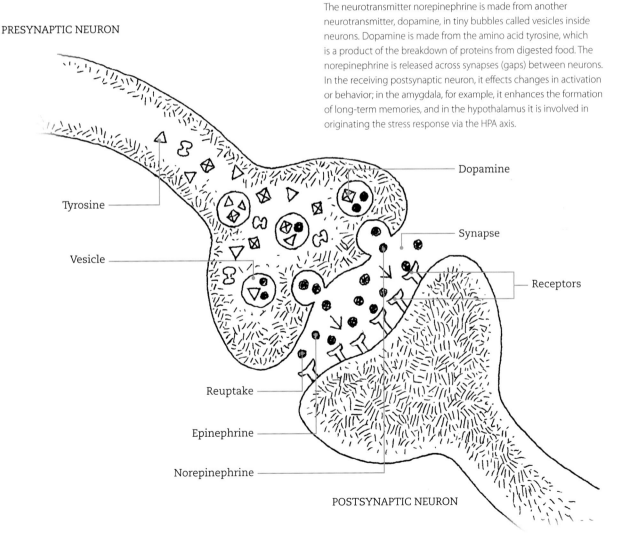

PRESYNAPTIC NEURON

Tyrosine

Vesicle

Reuptake

Epinephrine

Norepinephrine

Dopamine

Synapse

Receptors

POSTSYNAPTIC NEURON

TOP-DOWN MECHANISMS
The effects of psychosocial stress on the immune system

Top-down mechanisms are those initiated via mental processes at the level of the cerebral cortex. It has been proposed that the anterior cingulate cortex, prefrontal cortex, and the insular cortex are the principal brain areas involved. These cortical regions integrate information relevant to ongoing social, cognitive, emotional and personality issues, psychological stress, and challenges to maintaining equilibrium in internal organs, and send signals to distal parts of the body, including the heart, lungs, intestines, muscles, and even tissues.

Systems under stress
Many physical illnesses, including high blood pressure and peptic ulcers, appear to be partly related to stressors in everyday life. There is ample evidence of intercommunication between the central nervous system and the immune system under psychological stress. The central nervous system can influence the immune system in various ways. One of them is through monocyte-macrophages, which are the prime recipients of stress-related information and the key constituent responsible for initiating the immune response. When acute phase responses happen, such as infection, trauma, and inflammation, macrophages engage the immune system to destroy foreign agents, remove damaged tissues, and repair organ injuries. It has been postulated that this immune activation may cause HPA axis hyperactivity to increase the secretion of stress hormones such as glucocorticoids and adrenocorticotropic hormone (ACTH), which in turn will inhibit macrophage function and serve as a brake to the inflammatory process to restore the immune system to its pre-activation state.

Depression
Studies have established a link between the central nervous system, the immune system, and various psychiatric brain syndromes, including depression and chronic fatigue syndrome, which are examples of stress-related psychiatric disorders which impair mental alertness as part of their symptoms.

IMMUNE SYSTEM UNDER PSYCHOSOCIAL STRESS

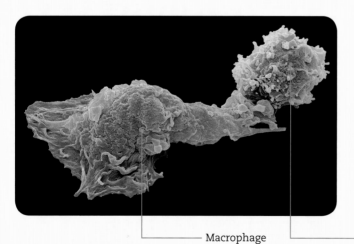

Macrophage   Th cell

MACROPHAGE
Colored scanning electron micrograph (SEM) showing the interaction between a macrophage (red) and a T helper lymphocyte (Th cell, blue), two components of the body's immune system. Both are types of white blood cell. Macrophages are antigen-presenting cells. They present antigens (fragments on the surface of pathogens or foreign objects) to T lymphocytes, activating them. Each T lymphocyte recognizes and binds to a specific antigen. Binding of the Th cell to the antigen presented by the macrophage activates the Th cell. This leads to its proliferation and the activation of other immune cells that eliminate the antigen. Stress hormones can inhibit macrophage activities and suppress our immune functions.

There is much literature suggesting that depression is caused or triggered by stress. The key features of depression are sadness and loss of interests. Decreased concentration and mental alertness are frequently found in patients with depression. In studying how stressful events may lead to depression, researchers have developed a theory called "learned helplessness." This is based on the observation that when animals experience prolonged or repeated stressful events that they are unable to change, they stop trying to escape from the adverse environment. Research has shown that the expression of helpless behaviors is associated with the amygdala area of the limbic system and the dorsal raphe nucleus in the brain stem. Depressed patients show similar learned helplessness behaviors, when they believe there is no control over the stressful situation based on perceived failures in the past. This helplessness frequently leads to low motivation for problem solving, which worsens the patient's ability to function.

The relationship between depression and the HPA axis has been a topic of interest in the past four decades. Many patients with depression fail to self-regulate and suppress blood cortisol levels when an external dose of cortisol is given, suggesting dysregulation of the HPA axis. Such a dysregulation is frequently found in patients with severe depression.

Chronic fatigue syndrome

Chronic fatigue syndrome (CFS) is another psychiatric disorder that impairs physical energy and mental alertness. The symptoms of this condition are pathological fatigue with at least four of the following features: memory problems, sore throat, tender lymph nodes, muscle and joint pain, headaches, unrefreshing sleep, and post-exercise muscle pain, in the absence of medical, alcohol/substance abuse, or psychiatric causes. Since most symptoms are also found in active viral infections, the etiology of the syndrome was first conceived as a viral infection. However, so far no single virus has been identified as the cause of chronic fatigue syndrome. A more consistent result was found in immune studies that showed decreased number and activity of NK cells and other T cell abnormalities.

While the immune abnormalities observed are fairly subtle, with inconsistent findings, the most consistent theme has been identified from cytokine studies, which showed that cytokines can produce chronic fatigue symptoms. For instance, high-level interleukin-2 (a type of cytokine) administration can cause fatigue, fever, and muscle pain. However, it is unclear whether immune changes are the primary cause of the syndrome, as patients who suffer from it also suffer from depression and anxiety disorders.

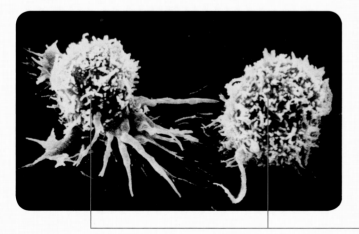

CANCER IMMUNOTHERAPY
Colored scanning electron micrograph (SEM) of two lymphokine-activated killer (LAK) cells, which have been activated by interleukin-2. LAK cells can kill cancer cells—but raised levels of cortisol produced as a result of stress reduce the proliferation of these cells. One approach to cancer therapy is to artificially culture these cells using interleukin-2 outside the body, and then reintroduce the cells into the patient from which they were taken.

Lymphokine-activated killer (LAK) cells

BOTTOM-UP MECHANISMS
The effects of physical distress on the brain

In complementing the concept of the top-down model, bottom-up mechanisms have also been proposed. Stimulation of various peripheral tissues and organs can influence brain information processing and mental activities via neural pathways that ascend from the different parts of the body to the brain stem and cerebral cortex.

The bidirectional mind–body connections

Connecting the bottom-up and top-down mechanisms gives the full picture of the bidirectional mind–body relationship. The proposed bottom-up mechanisms involve stimuli mediated by peripheral nerves associated with voluntary (somatic) and involuntary (autonomic, composed of sympathetic and parasympathetic) neural pathways. Somatic sensory nerves carry information from the voluntary part of the body, including the skin, striated muscles, and joints, while vagal nerves (parasympathetic) and spinal nerves (sympathetic) carry signals related to conditions within the involuntary part of the body such as the internal tissues to the brain. At the same time, top-down reciprocal nerves from the brain to the peripheral organs including the somatic and autonomic motor nerves provide the means by which the mind/brain influences bodily responses to ongoing situations or challenges. This forms an information feedback loop between various levels of the brain and peripheral tissues and organs to maintain equilibrium (homeostasis).

Optimal health depends upon the efficient exchange of information between the periphery organs and the brain and is closely associated with the vagal nerve. Communication along the vagal nerve pathways may be initiated, facilitated, or inhibited by hormones, immune-derived pro-inflammatory cytokines, and/or a variety of brain-derived neurotransmitters circulating in the body. Conversely, changes in mental processing (focused attention or perceived stress) generated in the brain can rapidly be expressed in the body via descending modulation of the autonomic, neuroimmune, and neuroendocrine systems.

The role of the immune system in these bidirectional mind–body connections

Immune and neural cells in response to both physical and psychological stressors secrete pro-inflammatory cytokines. They rapidly stimulate the brain to initiate immune activity in the periphery via multiple mechanisms, including the vagus nerve. Pro-inflammatory cytokines are also able to suppress response to glucocorticoids, and are potent stimulators of the HPA axis. These cytokines coordinate peripheral immune responses, which, in addition to promoting the creation of various blood cells and the release of various neurotransmitters, can also lead to the development of cardiovascular disease, diabetes, neurodegenerative disease, chronic pain, and depression. Peripheral and brain cytokines are also thought to play a critical role in cancer and autoimmune diseases.

Potential mechanism for the top-down and bottom-up models in mind–body therapies

Several interrelated mechanisms have been proposed to explain how the top-down and bottom-up models work hand in hand in mind–body therapies to improve homeostasis in the body. They include:

▸ Activation of specific brain structures, in particular the anterior cingulate cortex, prefrontal cortex, and the insular cortex, to improve balance of information between the left- and right-brain hemispheres;

▸ More efficient functioning of the limbic system and brain stem to fine-tune psychophysiological responses to enhance equilibrium in our internal organs;

▸ The regulation of the activation of certain genes (for example, growth factors, hormones) in our cells to respond to environmental stress.

It is considered that, through these proposed mechanisms, mind–body practices promote homeostasis via influence at multiple levels, from gene expression (cellular level) to the interaction of cortical brain regions that mediate systemic responses to internal and external challenges, including stress. It is likely that all mind–body practices actually involve a combination of both top-down and bottom-up mechanisms. Progressive muscle relaxation, for example, involves bottom-up neural pathways activated by various visceral activities (for example, reduced muscle tension, skin temperature, and blood pressure), which send signals to the brain; as well as top-down neural pathways activated by focused attention and the intention to relax, which send signals to joints and muscles to loosen up. Similarly, the practice of yoga, which incorporates meditation, breathing, and physical postures, combines the effects of peripheral input from bones, joints, muscles, and the respiratory organs, as well as the direct effects on the brain to induce both physiological and emotional changes in the body.

PARASYMPATHETIC AND SYMPATHETIC NERVOUS SYSTEMS
The autonomic nervous system controls automatic behaviors, such as changes in heart rate. It is divided into two subsystems that have opposite effects: the parasympathetic system is the rest-and-digest system, while the sympathetic system is the fight-or-flight response to stress. The systems largely work top-down, activated by signals from the brain. But signals can pass both ways, so information from tissues around the body has a bottom-up influence on our stress response.

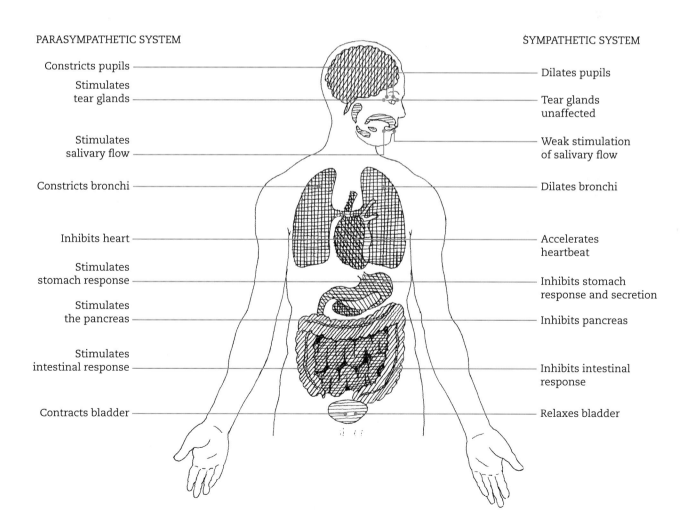

PARASYMPATHETIC SYSTEM

Constricts pupils
Stimulates tear glands
Stimulates salivary flow
Constricts bronchi
Inhibits heart
Stimulates stomach response
Stimulates the pancreas
Stimulates intestinal response
Contracts bladder

SYMPATHETIC SYSTEM

Dilates pupils
Tear glands unaffected
Weak stimulation of salivary flow
Dilates bronchi
Accelerates heartbeat
Inhibits stomach response and secretion
Inhibits pancreas
Inhibits intestinal response
Relaxes bladder

Chapter Five

STRESS AND THE SLEEP FACTOR

Stress is a natural reaction to difficult situations in life. Our reaction to stress can affect us on an emotional, cognitive, and biological level. Stressful experiences can often result in symptoms that make it difficult to relax in the evenings, and can prevent us from falling asleep. Insomnia and sleep issues are a common reaction to stressful experiences. Chronic stress due to an ongoing life situation is a common cause of sleep disorders. Stress activates our "fight-or-flight" response, which is designed to keep us alert, so it is easy to imagine how this system can interfere with sleep if it is chronically active. Stress can cause sleep disturbances and disturbed sleep can lead to more stress and an increased risk of a number of disorders. This can quickly become a vicious cycle that can be difficult to change without active intervention. In this chapter, we explore why sleep is so important to our well-being, the consequences of stress on our sleep patterns, and a number of strategies that can help with some of the more common sleep issues that individuals experience when they are stressed.

THE NEED FOR SLEEP
Can a good night's sleep reduce stress?

Sleep deprivation can make us more sensitive to stressful events, which can lead us to experience even more stress. The quality and amount of sleep we have can influence how we react to emotionally stressful events. One common effect of stressful events is the ability to affect the emotional content in our dreams. Dreams based on stressful events can sometimes be just as distressing as the original experience. Stress can also increase our startle response, decrease our ability to remember dreams, and change the quality of our sleep, often leading us to wake more often during the night. Stress also decreases our ability to experience REM (rapid eye movement) sleep, which is an important component of the sleep cycle. REM sleep is thought to be an important link in explaining how sleep affects our emotions the next day, although this relationship is still unclear.

BRAIN ACTIVITY DURING SLEEP

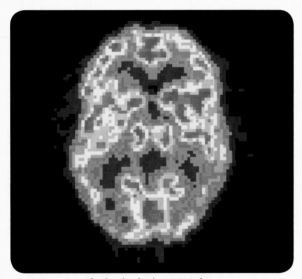

The brain during REM sleep

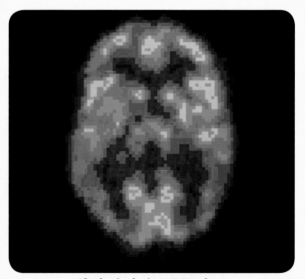

The brain during NREM sleep

THE BRAIN DURING REM AND NREM SLEEP
These images are colored positron emission tomography (PET) scans. The color-coded areas show the active cerebral brain in red through to the inactive parts in blue. During REM sleep (rapid eye movement)(left), the brain is active and dreaming, similar to being awake, note the red areas on the scan. During NREM (non-rapid eye movement) (right), the brain is inactive and sleeping deeply, note the blue areas on the scan, similar to being sleep deprived (see the image on page 89).

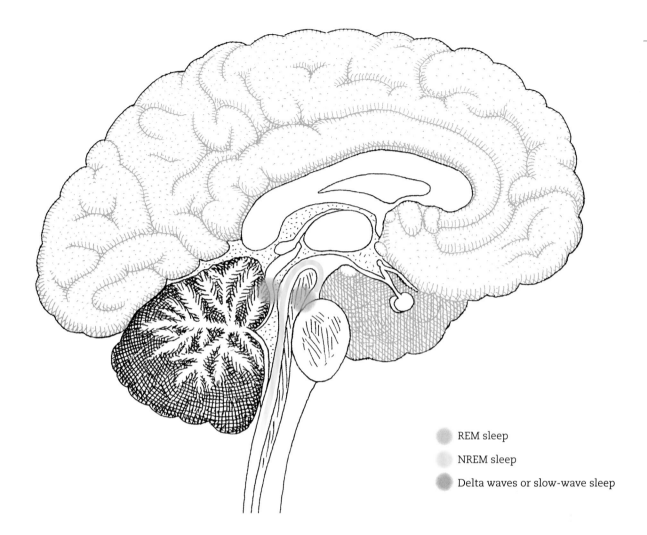

 REM sleep

 NREM sleep

 Delta waves or slow-wave sleep

The energy allocation theory

One well-supported theory of the purpose of sleep is energy allocation. This theory proposes that the purpose of sleep is to maximize our chances of surviving and reproducing by making sure we are using our energy in the most efficient way. One of the ways that sleep accomplishes this is by allowing an organism to "invest" in its biological systems. Our bodies require maintenance and upkeep in order to function well, and conducting this maintenance requires energy. We use a lot of energy while awake in order to move around and interact with the world. We use so much energy that we do not have much left over to repair and maintain the cells that make up our body. Sleep allows us to turn off, or at least turn down, some of our waking systems so that this energy can be used in the maintenance of our bodies.

THE BRAIN STEM SHOWING SLEEP STAGES
For REM sleep, neural activity originates in the brain stem. For NREM sleep, the cerebral cortex, thalamus, caudal brain stem, and spinal cord are involved. Delta waves, or slow-wave sleep, arise from either the cortex or the thalamus.

During REM sleep in mammals and birds, the temperature regulation defenses (which are largely used to regulate the temperature in response to increased activity, or environmental factors) and skeletal muscle tone are decreased dramatically, allowing our bodies to use this energy for cellular maintenance. By alternating REM sleep with NREM (non-rapid eye movement) sleep, we are able to save energy while maintaining our core body temperature and our ability to react to environmental changes.

The restoration of the plasticity of synapses

Energy savings and cellular repair are not the only reasons we need sleep. The synaptic homeostasis hypothesis (SHY) proposes that sleep is also the price the brain pays for plasticity. Plasticity is a term we use to describe the brain's ability to change and adapt to new circumstances. Our brain's cellular plasticity is the reason we are able to learn from our experiences. Our brains are very good at recognizing similar patterns and detecting changes in the environment. When we are awake, we make observations about our environment, and how it is the same or different from what we have seen or experienced in the past. This is done by strengthening cellular connections throughout the brain at the level of the synapse. As already noted, this increases our cellular energy needs.

In addition, the more we learn while awake, the more cellular "noise" there is. If we only strengthen the connections and increase the amount of cellular activity in the brain, eventually our ability to learn will become saturated, and we will no longer be able to distinguish meaningful information. So that we are able to distinguish meaningful information and irregularities in our environment, neurons need to fire sparsely and selectively. Sleep allows us to renormalize synaptic strength, restoring the ability of sensory neurons to prioritize more important stimuli, cellular homeostasis, and our ability to learn.

Causes of sleep deprivation

Many external and internal factors may contribute to sleep deprivation. Bright light during sleep time suppresses the production of the hormone melatonin, which plays an important role in the natural sleep-wake cycle. Acute and chronic pain are also sources of sleep deprivation. Sleep deprivation and pain have adverse effects on each other. Pain at night can lead to problems falling asleep and staying asleep, and poor sleep is linked to higher pain intensity during the day. Anxiety and stress have been shown to increase sleep arousal levels, leading people to wake up more often during the night.

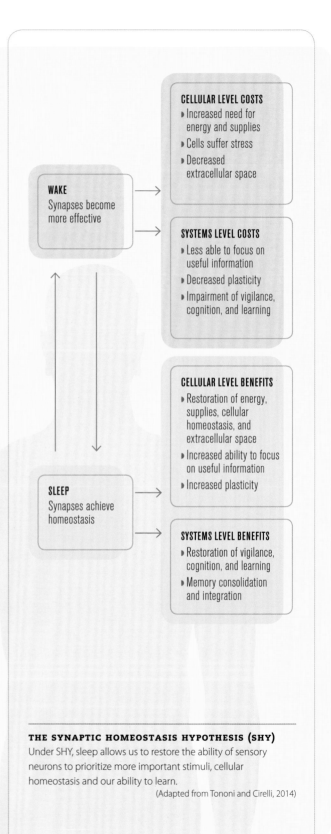

WAKE
Synapses become more effective

CELLULAR LEVEL COSTS
▸ Increased need for energy and supplies
▸ Cells suffer stress
▸ Decreased extracellular space

SYSTEMS LEVEL COSTS
▸ Less able to focus on useful information
▸ Decreased plasticity
▸ Impairment of vigilance, cognition, and learning

CELLULAR LEVEL BENEFITS
▸ Restoration of energy, supplies, cellular homeostasis, and extracellular space
▸ Increased ability to focus on useful information
▸ Increased plasticity

SLEEP
Synapses achieve homeostasis

SYSTEMS LEVEL BENEFITS
▸ Restoration of vigilance, cognition, and learning
▸ Memory consolidation and integration

THE SYNAPTIC HOMEOSTASIS HYPOTHESIS (SHY)
Under SHY, sleep allows us to restore the ability of sensory neurons to prioritize more important stimuli, cellular homeostasis and our ability to learn.
(Adapted from Tononi and Cirelli, 2014)

THE EFFECTS OF SLEEP DEPRIVATION ON THE BRAIN

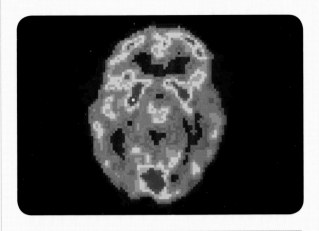

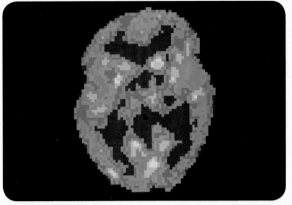

THE BRAIN WHEN AWAKE
A colored positron emission tomography (PET) scan of the human brain when it is awake. The color-coding depicts active cerebral brain areas (red) through to inactive areas (blue). When awake the brain is alert, with high brainwave activity, as can be seen in the red areas. The awake brain shows activity that resembles the phase of REM sleep when the brain is dreaming.

THE BRAIN DURING SLEEP DEPRIVATION
A colored positron emission tomography (PET) scan of the human brain during sleep deprivation. The color-coding shows active cerebral brain areas (red) through to inactive areas (blue). During sleep deprivation the brain shows general inactivity, similar to the phase of NREM deep sleep. Overtiredness leads to sluggish reactions and forgetfulness.

Consequences of sleep deprivation

The consequences of sleep deprivation are significant. Immune function can be altered, decreasing a person's ability to fight an illness. Sleep deprivation is positively correlated with slower recovery from illness. Studies have also shown that there is an increased risk of catching a cold after receiving less than seven hours of sleep per night. Sleep deprivation places the body in a perceived state of inflammation. Research has reported increases in inflammatory markers after subjects slept less than four hours, which leads to an increased immune system response, thereby stimulating the stress response.

When NREM sleep is fragmented or not achieved, patients can experience increased blood pressure and heart rate. Over time, this can lead to damage to the lining of blood vessels, which then release inflammatory markers, thereby further increasing cellular stress. Sleep deprivation has been shown to negatively affect metabolic and endocrine functions in a way similar to aging. As such, sleep deprivation may increase the severity of age-related chronic disorders. Sleep deprivation has negative effects on carbohydrate metabolism and if chronic, may contribute to the development of insulin resistance, leading to type 2 diabetes.

Sleep deprivation also results in decreases in cognition and alertness. Chronically sleep-deprived individuals' performance levels eventually stabilize at a lower level than before. Sleeping less than four hours per night has been shown to decrease brain activity by reducing the brain's metabolism of glucose. Since neurobiological processes involved in cognition are affected, sleep deprivation affects an individual's ability to make decisions. Sleep deprivation has also been correlated with an increased number of falls in hospital patients, since it can reduce reaction times, and may also affect neurological systems involved in balance. This is a major health risk, particularly for older patients.

THE CIRCADIAN RHYTHM
The importance of regular sleep

Evidence from controlled laboratory studies has linked circadian misalignment and sleep deprivation to the deregulation of the immune, inflammatory, and cardiovascular systems. The circadian rhythm plays an important role in the brain's capacity to maintain allostasis and reduce allostatic loading through homeostatic regulation of immune system activity. One night of total sleep deprivation does not seem to affect the level of most cytokines or monocytes in the blood, but immune cell rhythms are strongly regulated by sleep.

Sleep also plays a role in the elimination of free radicals. In our bodies, free radicals can interfere with normal cell functioning and cause damage to cell structures. They are usually formed from reactive molecules containing oxygen. These molecules are called reactive oxygen species (ROS). The damage caused by free radicals is known as oxidative stress. Anti-oxidative enzymes appear to rise and fall with the circadian rhythm and peak during the day. Although research into the effects of sleep deprivation on oxidative stress is limited, available data suggests that sleep deprivation and shifts in the circadian rhythm decrease the anti-oxidative enzymes in our body, which results in greater oxidative stress.

A potential mediator between circadian disruption and oxidative stress is melatonin. Melatonin is a hormone produced in the pineal gland, and its production is suppressed by light. It serves as a biological regulator, synchronizing the circadian rhythm with the light/dark cycle. Melatonin is also a powerful antioxidant. Individuals exposed to light late in the evening can experience chronic melatonin suppression, which leads to increased oxidative stress and further disruption of the circadian rhythm.

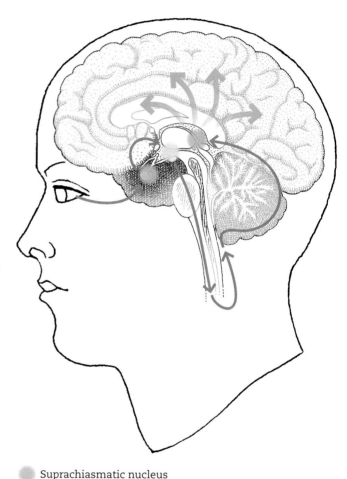

- 🔵 Suprachiasmatic nucleus
- ⚪ Paraventricular nucleus
- 🔵 Pineal gland
- → Melatonin

MELATONIN IN THE BRAIN
The suprachiasmatic nucleus receives inputs from photosensitive nerve cells in the retina and activates the paraventricular nucleus, in order to regulate the production of melatonin by the pineal gland. Lack of melatonin can disrupt the disrupt the sleep cycle.

Sleep and primitive stress patterns

Many hormones of the HPA axis play a role in influencing sleep, including corticotropin-releasing factor (CRF) and cortisol. Under stress, the CRF hormone increases our alertness and impairs sleep, and cortisol increases in response to stressful situations. Our baseline cortisol levels are regulated by the circadian rhythm, and can also be affected by the sleep/wake cycle, light, and exercise. Individuals who experience frequent disruption in these areas (for example, shift workers) have been shown to have chronically elevated cortisol levels.

Sleep inhibits the release of cortisol. In general, longer sleep is associated with better sleep quality, higher cortisol levels at awakening, and lower cortisol levels in the evening and while asleep. Sleep disturbances are associated with lower morning cortisol levels.

Individuals who get less sleep often need to exert more effort to get through the day. This increased effort can be stressful and may lead to increased cortisol levels, which remain high into the evening. Although the relationship between cortisol and sleep is still unclear, it has been hypothesized that high cortisol levels may be incompatible with sleep, and that sleep impairment is due in part to increased cortisol levels in the evening. By deregulating our cortisol levels, stressful events can lead to a self-reinforcing cycle of increased stress, higher evening cortisol levels, and sleep impairment.

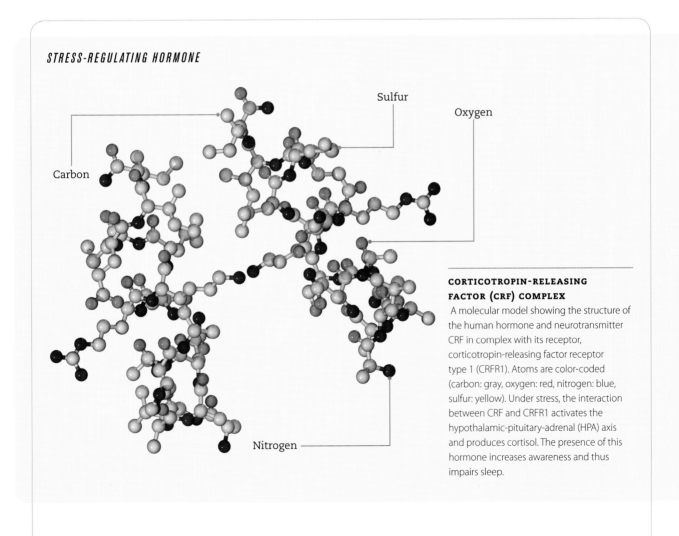

STRESS-REGULATING HORMONE

Sulfur

Oxygen

Carbon

Nitrogen

CORTICOTROPIN-RELEASING FACTOR (CRF) COMPLEX

A molecular model showing the structure of the human hormone and neurotransmitter CRF in complex with its receptor, corticotropin-releasing factor receptor type 1 (CRFR1). Atoms are color-coded (carbon: gray, oxygen: red, nitrogen: blue, sulfur: yellow). Under stress, the interaction between CRF and CRFR1 activates the hypothalamic-pituitary-adrenal (HPA) axis and produces cortisol. The presence of this hormone increases awareness and thus impairs sleep.

INCREASED ACTIVITY OF THE AMYGDALA
How we are stressed by a lack of sleep

Sleep likely also plays an important role in helping us process stressful experiences. Animal studies have shown that rats exposed to two hours of sustained stress have increased REM sleep in the 24 hours following the stressful experience. In the same study, the rats displayed increased activation of the hippocampus, and experienced changes in neuronal communication between the hippocampus and amygdala. These results suggest that a sustained stressful experience can lead to changes in sleep patterns and neuronal activity.

Sleep deprivation and the amygdala
Sleep-deprived individuals show increased activity in the amygdala, which is the emotional center of our brain. They also show weaker connections between the amygdala and the prefrontal cortex, which may lead to more uncontrolled emotional responses. This is supported by studies indicating that sleep-deprived adults take greater risks and are also less concerned about the future consequences of these risks.

In a recent study, researchers showed that the amygdala was much more active in a group of people who had missed a night's sleep than in the control group who slept a normal amount, when both groups were exposed to disturbing images. This suggests that when sleep deprived, even the brains of healthy people mimic certain pathological psychiatric patterns. The researchers concluded that sleep deprivation may contribute to depression and other psychiatric problems.

Sleep deprivation and the neural structure
Sleep deprivation can affect many of our core neural structures. One night of sleep deprivation decreases the blood flow to prefrontal areas of the brain the following day, and this decrease is associated with poorer performance on complex tasks. The prefrontal areas of the brain are particularly important for executive functioning and their ability to inhibit excessive stress-related amygdalar drive as described in Chapter Two. Young adults show fewer performance deficits following acute sleep deprivation, and are more able to adapt to chronic sleep deprivation compared to older adults. However, young adults are not immune from the long-term consequences of chronic sleep deprivation.

Sleep deprivation also decreases activity in a number of brain areas related to reward, including the caudate nucleus. This results in individuals needing greater stimulation in order to feel a sense of reward, and can lead to increases in impulsivity and reward-seeking behavior.

Medial prefrontal cortex (MPFC) area: keeps emotions and behaviors in check. Sleep-deprived individuals have weaker connections between the amygdala and the MPFC.

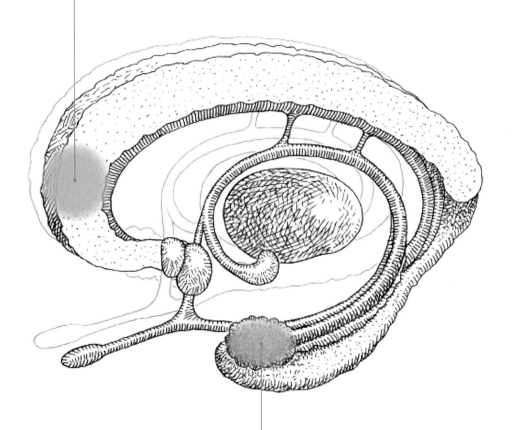

AMYGDALA ACTIVITY

A study comparing subjects with adequate sleep (sleep controls) to subjects who are sleep deprived under stress found that sleep controls had stronger functional brain connectivity between the amygdala and the medial prefrontal cortex. However, people with sleep deprivation showed stronger connectivity between amygdala and autonomic brain stem regions. The results suggest that under stress, people who are sleep deprived tend to have strong emotional reactions and less executive brain control of their emotions.

Amygdala: the emotional part of the brain, which is more active when deprived of sleep. This may explain overreaction and irrational emotions among sleep-deprived individuals.

LOSS OF REASON AND LOGIC

How sleep deprivation can put you at risk

A more recent avenue of research has investigated the effect of sleep deprivation on risk-taking behavior. Insomnia has been shown to be a causative factor for a number of risk-taking behaviors. In a twelve-month study of over 4,000 students from 7 to 12th grade, self-reported insomnia over the previous year predicted cigarette smoking and drunk driving, even after controlling for school grade, gender, and depressive symptoms. One striking study followed approximately 20,000 adolescents aged 10 to 19 years old and found that sleep disturbance, when combined with stressful negative life events, was associated with future aggressive acts such as carrying or using a gun. These effects remained after controlling for age, gender, and trauma-related stress.

Many large studies in the US and abroad have shown increased rates of substance use, drunk driving, unprotected sexual activity, violence and suicidal ideation in adolescents who get less than eight hours of sleep per night. One area of the brain that may be involved in this relationship is the ventromedial prefrontal cortex. This area of the brain is involved in complex decision-making and emotion regulation. Sleep-deprived individuals perform poorly on complex decision-making tests, and their scores are remarkably similar to patients who have sustained brain damage to the ventromedial prefrontal cortex.

Sleep deprivation in adolescents

Evidence also suggests that sleep plays a significant role in healthy adolescent development, especially in regulating daily functions related to behavior, emotion, and attention. Adolescents who do not get adequate amounts of sleep experience impairments in measures of health and daytime functioning. Experimentally induced sleep restriction in adolescents has been shown to produce declines in many neurocognitive functions. In addition

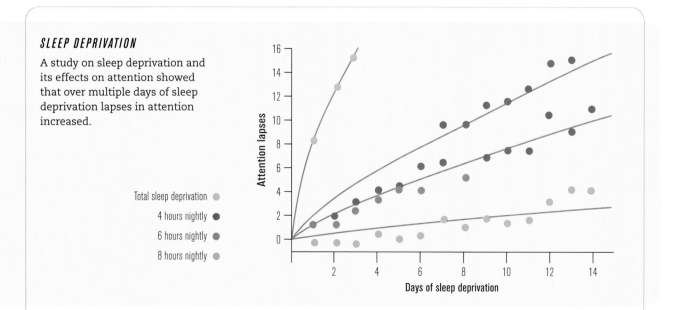

SLEEP DEPRIVATION

A study on sleep deprivation and its effects on attention showed that over multiple days of sleep deprivation lapses in attention increased.

Total sleep deprivation
4 hours nightly
6 hours nightly
8 hours nightly

Attention lapses

Days of sleep deprivation

SHIFT WORKERS
Shift workers are prone to sleep deprivation, or dysregulation of their circadian rhythms, which may lead them to undertake unnecessary risks or to behave unwisely.

to feeling tired following sleep restriction, adolescents experience cognitive deficits in complex functions such as verbal fluency and creativity, computational speed, and abstract problem solving.

However, adolescents did not experience deficits in low complexity tasks such as verbal memory, auditory attention, visual attention, psychomotor speed, or computational accuracy. Insomnia has also been shown to be a causative factor for a number of risk-taking behaviors.

Sleep deprivation in college students
Inadequate sleep is also a serious issue for college students. Fifty percent of college students report daytime sleepiness, compared to 36 percent of adolescents and adults; and 70.6 percent of college students report getting less than eight hours of sleep. Lack of sleep and irregular sleep schedules can negatively impact learning, memory, and cognitive

performance. Young adults are also more likely to drive while feeling drowsy or intoxicated. Sleep deprivation and drinking have been shown to have cumulative effects on driving impairment. Decreased sleep and irregular sleep schedules have been associated with increased depressive symptoms across a wide range of populations. Improving sleep has also been shown to improve depressive symptoms in college students.

Sleep deprivation in shift workers
Shift work is associated with irregular schedules, and workers often experience sleep deprivation and changes to their sleep cycle and circadian rhythm. The consequences of shift work include feeling tired and fatigued, which are often reversible after getting enough sleep, but can also include more severe long-term health risks such as increased risk for cardiovascular diseases and cancer. Shift workers also experience an increased amount of traffic and work-related accidents related to sleepiness. Evidence from controlled laboratory studies has linked circadian misalignment and sleep deprivation to the deregulation of the immune, inflammatory, and cardiovascular systems, which can ultimately result in illness.

SLEEP AND MEDICAL DISORDERS
How lack of sleep can lead to illness

Oxidative stress has been shown to play an important role in the aging process and the development of many diseases, including cancer. Sustained stress can lead to chronically elevated levels of reactive oxygen species (ROS), and the free radicals they produce. As noted in the last section, free radicals can interfere with cell functioning and cause damage to cell structures. Free radicals can modify and induce mutations in our DNA, resulting in altered cell function. They can also affect gene expression, and inhibit cell communication, sometimes resulting in increased cell growth, and a decrease in the normal cell death cycle. Increased levels of oxidative DNA damage have been noted in many types of cancerous tumors. Since sleep deprivation and the deregulation of our circadian rhythm have been shown to increase the amount of reactive oxygen species in our body, they may play a role in cancer development.

Obstructive sleep apnea may also increase the risk of cancer. Sleep apnea patients frequently experience hypoxia, a condition where our bodies do not receive enough oxygen. Hypoxia is also associated with various stages of tumor formation and progression, and may increase the amount of reactive oxygen species in our bodies.

Epidemiologic studies have found evidence of moderately increased rates of breast cancer in Caucasian female night and shift workers who have experienced repeated circadian disruption in their career. Further research is needed to extend these finding to other ethnic populations, since breast cancer has slightly different genetic profiles in other groups. Some studies have also shown increases in colorectal, prostate, endometrial, and other cancers, but research conducted to date has been limited and inconsistent. In 2007, in response to these findings and strong experimental animal research,

The World Health Organization International Agency for Research on Cancer classified the circadian disruption experienced by night and shift workers as "probably carcinogenic."

Neurological disorders

Sleep is important for the functioning of our nervous system, and is altered in many neurological disorders including dementia, movement disorders, epilepsy, headaches, demyelinating diseases, and cerebrovascular disorders. Patients with neurological disorders sometimes experience damage to areas of the brain that control sleep and produce pain. They may also experience motor disturbances and side effects from taking medication, which make it more difficult to sleep. In addition to being a common symptom, sleep deprivation can also increase the risk of developing neurological disorders, worsen other symptoms, and even interfere with treatment. The mechanism underlying increased rates of such disorders in sleep-deprived patients are complex, and likely involve an interaction between immune, neuroendocrine, autonomic, and vascular pathways.

Sleep loss increases the likelihood of hypertension, obesity, and type 2 diabetes, which are all elements of the stress-related metabolic syndrome we introduced in earlier chapters. The metabolic syndrome increases the risk for heart disease and stroke as well as diabetes. Obesity is associated with obstructive sleep apnea, which interferes with sleep and increases the risk for stroke. One large study of over 5,000 participants found that individuals who had less than six hours of sleep per night had a fourfold greater risk of stroke than those who had seven or eight hours of sleep. The link between stroke and sleep deprivation is still not well understood. It is thought that this increased risk of stroke may be due to chronic stress, which increases cortisol levels and heart rate.

Interestingly, a number of studies have found that excessive sleep increases the risk of developing Parkinson's disease. The reasons for this are still unclear, but may be due to abnormalities in Parkinson's patients that predate their diagnosis by many years.

In seizure disorders, sleep deprivation increases seizure frequency. Additionally, patients with obstructive sleep apnea have poorer control of seizures, and the most common treatment for this condition (a continuous positive air pressure device or CPAP) allows for better control of seizures.

Sleep deprivation has also been identified as a factor that often precedes migraines. Therefore, people who get less sleep may experience more frequent migraines.

Hypertension

Sleep deprivation is associated with increased blood pressure and an increased risk of hypertension. Sleep duration of less than five to six hours per night increases the risk of developing hypertension in individuals under sixty years of age, and persistent insomnia is associated with an increased risk for hypertension in middle-aged individuals. Sleep loss may act as a chronic stressor that activates our nervous system and leads to sustained increases in blood pressure. Chronic inflammation associated with sleep deprivation is also thought to play an important role in an increased risk for hypertension.

Cardiovascular disease

Inflammation has been shown to play a key role in the development and progression of cardiovascular diseases. The development of plaques in our blood vessels is a complex inflammatory response to stimuli in our blood stream. Sources of system-wide inflammation, such as arthritis, can accelerate the development of these plaques by increasing the overall level of inflammation in our bodies. Sleep deprivation increases many markers of inflammation, and has also been associated with an increased risk for many cardiovascular diseases.

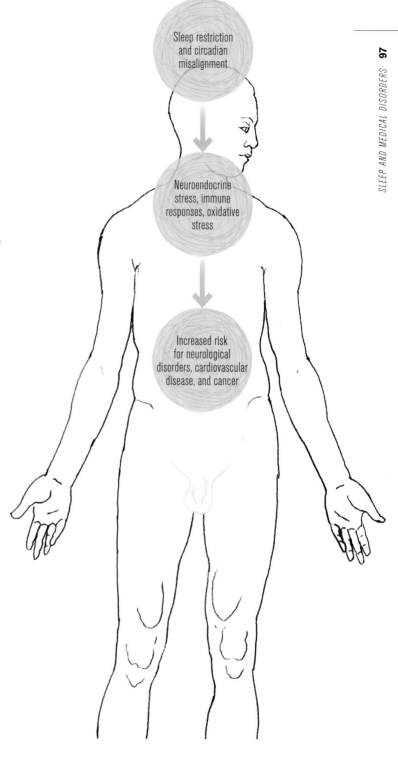

SLEEP DISTURBANCE AND RISK OF DISEASE
Sleep dysfunction has been found associated with higher risks for various medical disorders including cancers, neurological disorders, and cardiovascular diseases.

SLEEP AND MENTAL WELL-BEING

How sleeps affects psychiatric and cognitive disorders

Sleep plays a critical role in emotion regulation, and a lack of sleep may worsen symptoms of existing psychiatric disorders or increase the risk of developing psychiatric disorders in the future. Insomnia is a common symptom in many psychiatric disorders, and especially mood and anxiety disorders. Sleep-deprived individuals tend to remember more negative experiences and fewer positive ones. Sleep allows us to restructure our experiences and memories in an emotionally adaptive way. Studies using functional magnetic resonance imaging (fMRI) have shown that sleep deprivation causes alterations in many of our emotional brain networks, including the limbic system.

Excessive sleep may also play a role in major depression. Too much sleep has been associated with decreased mood, impaired functioning, and a number of negative health outcomes. Excessive sleep can be indicative of a major depressive episode, and some studies have shown that controlled sleep restriction can improve symptoms in depressed patients.

A number of neurobehavioral problems in children have been associated with sleep deprivation. Persistent sleep deprivation in children between two and a half years old and six years old predicts high hyperactivity and impulsivity scores, and poor performance on neurodevelopmental tests when these children enter school between the ages of five and six. While this result is compelling, it is not yet clear whether sleep deprivation, hyperactivity, or a combination of both is the main cause.

Sleep problems in childhood also predict decreased general executive functioning at sixteen years old. These findings suggest that there may be a critical development period when sleep disruption can result in the development of issues later in life. Sleep problems are also very common in both attention

deficit hyperactivity disorder (ADHD) and autism spectrum disorders, and sleep deprivation is thought to play an important role in these disorders.

The mechanism underlying neurobehavioral dysfunction following sleep deprivation is still not well understood. Evidence suggests that it is likely a combination of dysregulation of the circadian rhythm and the hormones and neurotransmitters involved with alertness.

Cognitive impairment

Sleep deprivation can also affect cognitive performance. Cognitive performance is a broad concept that includes a number of distinct processes including memory function, the detection of changes in our environment, the coordination of body movements, and many other domains. There is significant individual variation in how sleep deprivation affects different areas of cognitive performance, and some tasks may rely more on certain cognitive processes than others. The reason sleep deprivation has different effects on certain areas of cognitive functioning is poorly understood, and further research is needed.

Animal studies have shown that sleep deprivation affects the structure and function of the hippocampus, which plays an important role in memory. Studies in older human adults have found that increased levels of cortisol are associated with greater memory impairments and lower hippocampus volume. This points to a potential relationship with sleep deprivation, which has been shown to increase cortisol levels. It is possible that sleep deprivation in humans decreases the ability of our hippocampus to produce new neurons. This would decrease our ability to form new, adaptive memories and have negative consequences on cognitive performance.

Experimental studies have demonstrated that getting less than seven hours of sleep per night can cause many problems. These include deficits in alertness, attention, cognitive speed, and memory. Studies have also found that the effects of chronic sleep deprivation on cognitive performance cannot generally be reversed by just one night of recovery sleep. Individuals who sleep more hours per night prior to sleep restriction experience less severe deficits, and also reverse these deficits faster following repeated nights of recovery sleep.

HYPERACTIVITY IN CHILDREN
Sleep dysfunction may play an important role in attention deficit hyperactivity disorder (ADHD) and autism spectrum disorders.

THE POWER NAP

Why are power naps helpful and what do they do?

Napping is one potential way to minimize the neuroendocrine and immune stress associated with sleep deprivation. Naps of at least thirty minutes have been shown to decrease cortisol levels in human subjects, and napping may relieve stress by inhibiting the HPA axis and cortisol release. One study found that following a night of acute sleep deprivation where subjects only got two hours of sleep, a thirty minute midday nap improved alertness. Furthermore, only subjects who took a midday nap experienced a reduction in immune system inflammatory markers to normal levels after a night

of recovery sleep. This suggests that a standard eight hours of recovery sleep is not enough to normalize immune alterations after sleep deprivation unless accompanied by a midday nap.

Historically, in southern European countries a "siesta" or nap is taken after lunch. A study of 23,681 individuals in Greece, where midday naps are relatively common, compared to the United States, reported that midday napping in healthy, working men was associated with significantly less heart-related deaths, even after controlling for potential confounders. In Japan, too, office workers are now encouraged to take naps during their working day.

Many studies have found that napping may be helpful during night work shifts. Night work shifts have been shown to significantly disrupt the circadian rhythm, and napping may help to protect from some of these negative effects. Planned naps during night shifts reduce sleepiness and improve cognitive performance deficits associated with sleep deprivation. Naps as short as twenty minutes can have beneficial effects. However, there is also generally a period of twenty to thirty minutes following sleep where it is common to experience decreased cognitive functioning. This is known as sleep inertia. Not surprisingly, studies conducted in shift workers also found that a single nap is not enough to fully recover from working all night. That is, naps should not be used as a substitute for a full night of sleep.

Fatigue and cognitive performance due to sleep deprivation are a particular concern for pilots. Sleep deprivation can impair a pilot's ability to recognize and effectively respond to problems while in flight, and pilot fatigue has been identified as a contributing or causal factor in many aviation accidents. Napping is one strategy that has been successfully implemented in studies conducted in pilots. One study found that pilots who had an opportunity to nap for forty-five minutes were more alert during the approach and landing phases of flight, which is when most aviation accidents happen. Naps can maintain and restore alertness and performance, and strategic scheduling of naps between copilots can avoid the potential negative effects of sleep inertia on safety. More research is needed in order to measure the effects of napping on specific flight performance and safety measures.

Naps may also help with academic performance in adolescents and young adults. In one survey, high academic performers were more likely to nap than low academic performers (52 percent versus 29 percent, respectively). Another study which tested students on a face and name recognition task found that subjects performed significantly worse after 6 pm, except for those who had experienced a 100-minute nap during the day. The napping group also showed improvement on this task when retested later on.

In general, sustained wakefulness can impair cognitive performance and increase overall stress, and napping is a useful strategy to counteract these effects. It is not always possible to take naps when we need them most. It can be difficult to fall asleep during periods of stress, and many settings, particularly the workplace, do not have quiet spaces appropriate for napping. However, even naps in less than ideal settings may have beneficial effects. Further research is needed to examine the effect of naps on safety outcomes in the workplace. The timing of naps may be important, especially considering the possible effects of sleep inertia on important tasks, which might endanger life, following napping.

NAPPING

Having a short nap may decrease the neuroendocrine and immune stress associated with sleep deprivation, and counteract the effects of the HPA axis triggered by stress.

THE SLEEP HYGIENE ROUTINE
How to get a good night's sleep

There are many commonly held beliefs about sleep that may contribute to insomnia. One commonly held belief is that we require eight hours of sleep to function during the day. Much of the research we have outlined in this chapter has used eight hours as a benchmark for a healthy amount of sleep. However, there is a wide range of sleep needs in adults. Some individuals may be able to function well with less than eight hours of sleep per night, while others may need more. While sleep deprivation often results in decreased cognitive functioning, it is still possible that some are still capable of functioning after less than eight hours of sleep. Sleep duration is only one factor in how well we function the next day. Sleep quality is likely to be just as important as the amount of sleep we get, and anxiety related to sleep duration can actually prevent us from getting a significant amount of good quality sleep.

Another common belief about sleep is that you should always wake up feeling refreshed. As we saw in research studies investigating napping, it is natural to experience some sleep inertia after waking. You may feel groggy and have decreased cognitive functioning for a period of time. The degree of sleep inertia you experience may depend on the sleep stage from which you awake, and other factors which are still not well understood. Sleep inertia is not necessarily an indicator of not getting enough sleep, and, furthermore, anxiety surrounding this issue can contribute to insomnia.

It is also believed that waking up periodically during the night means that sleep is not restful. Brief awakenings are actually a normal part of the sleep process, and do not by themselves indicate poor sleep.

Cognitive behavioral therapy for insomnia
Cognitive behavioral therapy (CBT) is an effective treatment for chronic insomnia. The goal of CBT for Insomnia (CBT-I) is to directly target and change beliefs, thoughts, and behaviors that make it difficult to sleep. Some of the factors that CBT-I targets include spending too much time in bed, excessive daytime napping, irregular sleep schedules, rumination over the consequences of poor sleep, and anxiety experienced when trying to fall asleep. CBT-I treatment generally consists of four to ten sessions of fifty to sixty minutes. Sessions occur approximately once per week, and there is usually a two-week period before beginning treatment where patients are asked to self-monitor their sleep-related behaviors.

The behavioral component of the treatment attempts to reestablish an association between the sleep environment and sleep. Patients are asked to make the following changes to their sleeping routine:
▶ Only use their bed for sleep, and not for any other activities such as reading or watching TV.
▶ Limit the time they spend in bed to only time that is actually spent sleeping.
▶ Select a standard wake-up time and then select a bedtime that only includes the number of hours actually spent sleeping on an average night. (This change can actually lead to getting less sleep in the short term and individuals with bipolar disorder should only attempt this under the supervision of a mental health professional.)

For the cognitive component of the treatment, the therapist will address and attempt to modify cognitive processes that perpetuate insomnia. The therapist may ask patients to record their thoughts about sleep each night and conduct behavioral experiments in order to modify thoughts and beliefs that contribute to insomnia.

The therapist will also provide instructions on proper sleep hygiene, and will outline external factors that may impact insomnia. Common factors that can affect our ability to fall asleep include drinking caffeine and alcohol in the evening. Caffeine increases the amount of time it takes to fall asleep. Alcohol's sedative effects can make it easier to fall asleep, but worsen sleep quality in large quantities. It is also important to get regular exercise. Many sleep specialists consider exercise an effective intervention for sleep problems. However, exercise in the late evenings can actually interfere with sleep.

Sleep hygiene strategies alone are generally not enough to treat insomnia, but are still an important part of treatment. Effectiveness studies of CBT-I have found that none of the individual components are as effective as full CBT-I. Full CBT-I is associated with the most improvement in insomnia symptoms, which were maintained six months after treatment. CBT-I is the most recommended treatment for insomnia, and studies suggest that 70 to 80 percent of patients will experience a reduction in insomnia, and 40 percent will experience a complete remission of insomnia. One study also found that CBT-I significantly

SLEEP HYGIENE
For people who suffer from insomnia, it is essential to follow a sleep routine to encourage a deep and restful sleep. If insomnia persists despite following sleep hygiene measures, additional types of treatment may be needed. *Sleeping Beauty* by Edward Burne-Jones, 1874.

decreased the use of sleep medication. If individual CBT therapy is not possible, group and self-help CBT treatment options are also available.

Other treatments
There are many medications that have shown some effectiveness in treating insomnia. You should consult with your doctor if you are interested in trying a particular medication.

More recently, some research has investigated mindfulness-based therapies as a treatment for insomnia, which incorporates mindfulness meditation practices with behavioral strategies. Early results have been promising, showing that combining mindfulness meditation with CBT-I leads to reductions in symptoms that remain twelve months after treatment.

Chapter Six

STRESS AND WOMEN'S HEALTH

Why devote an entire chapter to the topic of stress and women's health? Men and women have different amounts of reproductive hormones that affect much more than our physical appearance and sexual health. These reproductive hormones affect nearly every organ in our bodies including our major stress response system, the hypothalamic-pituitary-adrenal (HPA) axis. Although men were previously considered at increased risk for stress-related diseases compared with women, this gender gap no longer applies. In fact, women now outnumber men in terms of autoimmune disease, anxiety, and depression. Particularly during their reproductive years, women appear to be at increased risk for stress-related disorders. Although social factors play a role, including the significant demands of balancing work and motherhood, it appears that changing levels of female hormones—both across the menstrual cycle and across the lifespan—directly impact a woman's vulnerability to stress. This chapter will highlight the interactions between some key female reproductive hormones and stress. Men should also take note, however, since these hormones are also produced in men, albeit at different and generally steadier amounts.

OXYTOCIN
The anti-stress hormone

Known as the "love hormone" or "cuddle hormone," oxytocin is produced by the brain's hypothalamus and is known for its important role in reproduction and mother–infant attachment. It is also increasingly appreciated as a stress-regulating hormone whose neurons project to important regions in both the male and female brain.

Levels of this feel-good hormone increase with wide-ranging activities including childbirth, breastfeeding, sex, and simply thinking fondly of a person. Although it was initially branded as a universal anti-stress, feel-good hormone, oxytocin is increasingly appreciated as a more complex hormone that can have paradoxical, context- and person-dependent effects. In general, however, oxytocin is associated with social attachment behaviors that play an important role in offsetting stress.

It is now well accepted by the medical community that positive social interactions and social support are good for your health and longevity. Among the known health benefits of social support are decreased cardiovascular reactivity, improved mood, and decreased cortisol levels in response to acute stress.

Some believe that social attachments may be especially beneficial for reducing stress in women. This makes intuitive sense if you consider that the typical response to stress we have emphasized throughout this book, the epinephrine- or adrenaline-based fight-or-flight response to stress, may not be particularly advantageous to a woman at different points in her life, particularly pregnancy or at other times of being "encumbered" by a dependent child. From an evolutionary standpoint, a woman is better off forming affiliations with others who could warn her about danger, help her flee, or help care for her baby. This is known as the "tend–mend–defend" response. In contrast, men are naturally more inclined to an aggressive "fighting" response to stress thanks to their higher testosterone levels, relatively increased muscle mass, and absence of an "encumbered" pregnancy state. This is relevant to a stressed-out woman since her brain is more naturally geared toward an affiliative response to stress compared with her male counterparts. Women, when stressed, are more inclined to seek out others for comfort. Men, on the other hand, are not as inclined to react in such a "touchy feely" way when stressed. This difference is largely driven by the fact that male and female brains are wired for oxytocin differently.

Regardless of your gender, though, oxytocin leads to a number of positive, homeostasis-restoring changes. Oxytocin can make us feel better by: releasing dopamine (an important brain chemical) in our brain's pleasure centers; lowering cortisol; increasing natural pain-reducing chemicals; and decreasing the firing of certain "fight-or-flight" neurons. Oxytocin has also been shown to boost levels of our brain's sedative, gamma-aminobutyric acid (GABA), which helps us return to a resting, calm state. Oxytocin can also serve as an "on or off" switch for different stress-related genes, thereby influencing the structure and function of important regions of our brain. This latter aspect may explain why oxytocin levels early in life (including infancy) can lead to long-lasting changes in how we react to stress.

FEMALE HORMONE SYSTEM

This system has hormones with separate but interrelated and wide-reaching functions. Oxytocin is a key hormone secreted by the pituitary gland in the brain with important reproductive and social attachment functions. Endorphins are "feel-good" hormones secreted by organs that increase your pain threshold and are produced with exercise and other activities. Vasopressin regulates the body's retention of water and constricts blood vessels. Like oxytocin, it also appears to regulate social cognition and behavior. Estrogen and progesterone are hormones that have important reproductive and social functions.

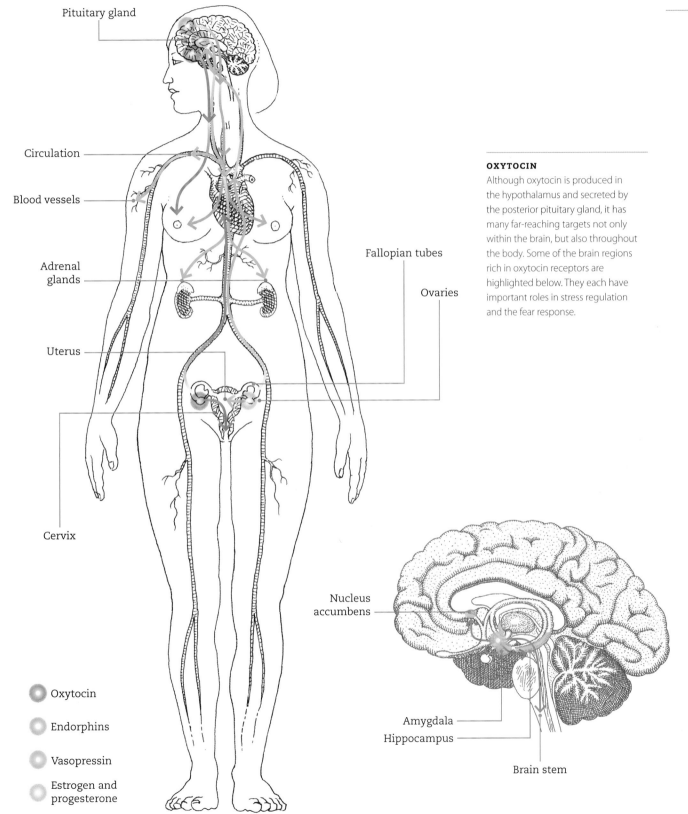

Pituitary gland

Circulation

Blood vessels

Adrenal glands

Fallopian tubes

Ovaries

Uterus

Cervix

OXYTOCIN

Although oxytocin is produced in the hypothalamus and secreted by the posterior pituitary gland, it has many far-reaching targets not only within the brain, but also throughout the body. Some of the brain regions rich in oxytocin receptors are highlighted below. They each have important roles in stress regulation and the fear response.

Nucleus accumbens

Amygdala

Hippocampus

Brain stem

- Oxytocin
- Endorphins
- Vasopressin
- Estrogen and progesterone

Oxytocin and the stress response

However, the relationship of oxytocin to stress is not as clear-cut as initially believed. In certain contexts, oxytocin may actually increase your stress response. This paradox is partly explained by the fact that oxytocin appears to increase "social cue salience." In other words, it helps you to read other people's emotions better. Depending on the context and whether the environment is seen as safe or threatening, as well as your own psychological vulnerabilities, you will either be more stressed or relaxed in the company of another person. Social support with a familiar, non-threatening person, such as a family member, will lower your stress levels. On the other hand, if the social support comes from an unfamiliar person, the stress-lowering properties of oxytocin disappear. This context-dependency makes sense if you remember what we just mentioned about oxytocin causing people to pay more attention to social cues. In the presence of a stranger, your "social hormone" oxytocin is alerting you to a person's unfamiliar habits or expressions, leading to increased vigilance and relatively increased stress levels, compared with if you were in the company of someone familiar and presumably non-threatening, whose facial expressions you did not have to heed with such alertness.

Different levels of oxytocin

Another nuance of oxytocin is that we all differ in our baseline natural levels of it. For example, individuals with Asperger syndrome and other autism spectrum disorders have naturally lower oxytocin levels. This is not surprising if you consider that a central feature of autistic people is their difficulty in dealing with social situations. Interestingly, when people with autism are given oxytocin, their ability to read other people's emotions improves.

Finally, the nature of the oxytocin exposure itself can impact its ability to lower stress hormone levels. In general, the more frequently you get it, the better for you. Giving people repeated doses of oxytocin has been shown to lower their blood pressure and cortisol levels while increasing their levels of natural opioids (pain-dulling, feel-good chemicals). Release of norepinephrine (a major fight-or-flight hormone) is also blocked, leading to decreased stress reactivity.

THE BENEFITS OF OXYTOCIN

STRESS-BUFFERING HORMONE
A polarized light micrograph (PLM) of crystals of oxytocin. Oxytocin is a peptide of nine amino acids important for social attachment and reproductive functions including orgasm, labor, and milk let-down. The name is derived from the Greek *oxys tokos*, meaning swift birth. Oxytocin is increasingly appreciated as a stress-buffering hormone, but its specific effects on one's stress response appear to be context-, dose-, and person-dependent.

How to boost oxytocin

Fortunately, we don't have to go to a physician or a lab to increase our oxytocin levels. There are many strategies we can turn to during times of stress to self soothe and boost oxytocin naturally. These include wide-ranging activities such as sex, massage (no surprise that health benefits of regular massage therapy exist), eating (which can explain the all-too-frequent phenomenon of stress eating), cuddling with or hugging a loved one, exercising, breastfeeding, getting enough of the "sunshine" vitamin D, and even petting a dog.

Higher oxytocin levels in dog owners might help explain the increasingly popular health-care trend of physician prescriptions for therapy dogs to help treat anxiety, depression and cardiovascular disease. "Oxytocinized" dog owners have better health outcomes after surviving a heart attack than non-dog owners. If we go back to the concept described earlier of the importance of regular, repeated doses of oxytocin for optimal stress-lowering, dog ownership further illustrates this point; petting or playing with your dog presumably happens with enough frequency (for example, multiple times a day) to boost oxytocin levels in the most beneficial way. Breastfeeding is another great example, since most breastfeeding

MOTHER-INFANT BONDING: A HIGH OXYTOCIN STATE
Oxytocin levels increase during physical contact with a loved one, including skin-to-skin contact between a mother and her baby. The high levels can result in health benefits for both mother and child.

mothers nurse multiple times a day. Unsurprisingly, studies of breastfeeding mothers show that they have a lower risk of developing type 2 diabetes and cardiovascular disease. In addition, the more they breastfeed, the less likely they are to develop these problems. This provides additional support that longer-term, repeated oxytocin exposure provides the best anti-stress and health-promoting benefits.

If natural oxytocin-boosting strategies are not enough, certain serotonin-based antidepressants (for example, Prozac and other serotonin reuptake inhibitors) have also been shown to increase oxytocin levels. Indeed, some believe that this little-known oxytocin-boosting property of serotonergic antidepressants accounts for their effectiveness. Like oxytocin, serotonin is another important natural brain chemical that helps you feel less stressed and anxious, so it is not very surprising that oxytocin and serotonin are interconnected in some way.

STRESS AND PREGNANCY
Will stress levels affect my baby?

Although we will read more about the detrimental effects of early life stress in Chapter Eight, one of the earliest life stressors is now the focus of this section: prenatal stress, or the effects of a pregnant mother's stress levels on her growing baby. Although stress hormones are necessary for the proper growth of a baby's organs and tissues during pregnancy, chronic or severe stress in the mother activates a chain of reactions that result in harmful physical changes in her baby. As we see throughout this book, it is the chronic, unremitting stress that is most detrimental to our health—and stress in pregnancy is no exception.

Prenatal stress

First, stress can activate genes that lead to increased leakiness of the placenta, which is the major organ connecting the growing baby to her mother. This paves the way for baby to be exposed to more maternal stress hormones (which pass across the placenta) as well as to elevated markers of inflammation (those "proinflammatory cytokines" that reappear throughout this book). This inflammatory state leads to more leakiness, and so on and so forth, creating a vicious cycle. The mother, placenta, and her baby are in this way primed to have an exaggerated or over-reactive immune response to any other potential issue that may come along during the course of the pregnancy. And exaggerated immune responses, much like exaggerated stress responses, as we see throughout this book, are generally not a good thing.

In the case of pregnancy, an exaggerated immune response to potential triggers (including additional psychological stress or dietary deficiency, low oxygen, infection, or other physical stressors) can result in damage to the placenta and to the baby's developing brain and other organs. These changes are genetically mediated. The particular timing and intensity of the immune response, combined with other factors such as the gender of the offspring—with male babies being extra vulnerable—is thought to drive the end result. In a perfect storm, structural brain changes can take place, leading to long-term emotional and cognitive effects, which include increased stress reactivity in the offspring.

PREGNANCY

Stress management is particularly important during pregnancy, a crucial period in a mother's and her baby's life. Stress during pregnancy, depending on its intensity or duration, can lead to exaggerated immune responses in the placenta and in the baby's developing brain. This can ultimately affect the baby's brain structure, function, and stress reactivity.

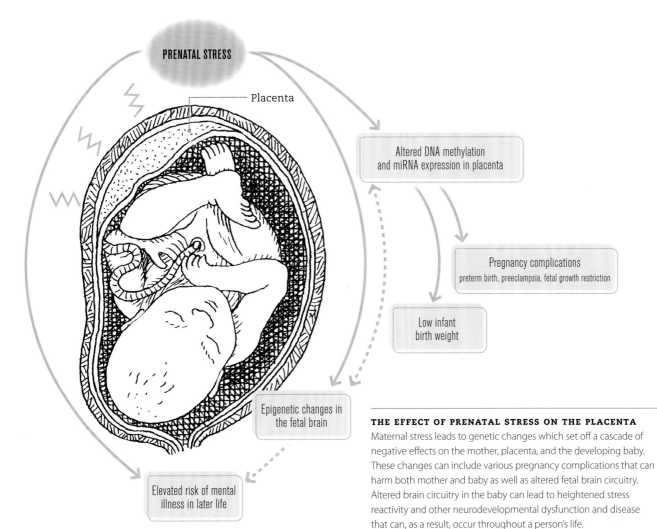

PRENATAL STRESS

Placenta

Altered DNA methylation
and miRNA expression in placenta

Pregnancy complications
preterm birth, preeclampsia, fetal growth restriction

Low infant
birth weight

Epigenetic changes in
the fetal brain

Elevated risk of mental
illness in later life

THE EFFECT OF PRENATAL STRESS ON THE PLACENTA
Maternal stress leads to genetic changes which set off a cascade of
negative effects on the mother, placenta, and the developing baby.
These changes can include various pregnancy complications that can
harm both mother and baby as well as altered fetal brain circuitry.
Altered brain circuitry in the baby can lead to heightened stress
reactivity and other neurodevelopmental dysfunction and disease
that can, as a result, occur throughout a person's life.

Inherited prenatal stress

Interestingly, stress-mediated genetic changes appear
to be inherited across several generations. Thus, if
your great-grandmother was subject to high levels
of stress while pregnant with your grandmother, this
alone increases your odds of developing stress-related
disorders thanks to inherited genetic influences.
Although skeptics might argue that a child of a
stressed-out parent might simply be learning this
parenting behavior and having it "transmitted"
this way, there is evidence of transgenerational
transmission of stress responses that occur even
after in vitro fertilization and raising the offspring in
a foster environment. So it seems that stress-induced
genetic changes are in fact transmitted to the next
generation via the sperm and egg.

Coping with prenatal stress

So what is a stressed out mother-to-be to do?
Fortunately, even if she did nothing, she'd be partly
protected thanks to naturally rising levels of certain
pregnancy hormones that are indirectly able to
ward off stress. First, estrogen levels increase during
pregnancy, and estrogen has been shown to increase
serotonin levels (the feel-good brain chemical
mentioned earlier, which is the target of action of the
most commonly prescribed class of antidepressants).
Intriguingly, female babies are thought to be better
protected against neurodevelopmental disorders like
autism and ADHD in part thanks to the fact that they
have more serotonin-boosting estrogen compared
with male babies, and serotonin plays a role in
guiding proper brain development.

THE PROPERTIES OF PROGESTERONE

PROGESTERONE HORMONE
A polarized light micrograph of crystals of the female sex hormone progesterone. Although progesterone was initially thought to serve primarily reproductive functions (much like oxytocin), it is increasingly understood to have stress buffering and anxiety-lowering properties, thus helping to reduce stress.

Progesterone helps to combat stress

Progesterone is another important reproductive hormone that is needed for pregnancy and that can also combat stress. High progesterone levels in pregnancy help calm the mother-to-be by increasing a natural sedative in her brain: gamma-aminobutyric acid (GABA), our major inhibitory neurotransmitter. GABA is like Mother Nature's Valium and is high during resting and relaxed states. It increases during meditation and yoga and may account for the phenomenon of "mommy brain" or brain fogginess during pregnancy. In addition to raising levels of GABA, progesterone also appears to activate a powerful anti-inflammatory gene that protects mom's cells against stress. Progesterone, then, seems to be her ally against inflammation and stress. Indeed, it may even help a woman's odds of having improved surgical outcomes. Several cancer studies show that women at higher-progesterone points in their menstrual cycle have better surgical outcomes compared with women in low-progesterone phases of their cycle. All things considered, it seems a good thing that progesterone levels increase throughout pregnancy—it's Mother Nature's way of optimizing the health of mother and baby!

Strategies to minimize stress

However, sometimes stress can get the better of even the most well-intentioned expectant mother. When stress levels overwhelm her built-in stress-lowering defenses, there are many different strategies she can use to minimize stress during pregnancy or to at least mitigate its harmful effects. These include meditation, yoga and other low-intensity exercise, ensuring adequate sleep, and proper nutrition.

Stress-fighting nutritional practices that may benefit both mother and her baby include consumption of low-mercury fatty fish (sardines, salmon), vitamin D (which boosts brain levels of calming serotonin), and tryptophan-rich foods (fish, chicken, eggs, nuts, and beans). Tryptophan is a dietary source of serotonin, and depletion of it leads to increased anxiety and depressed mood. Because low-mercury fish is a source of vitamin D, omega-3 fatty acids, and tryptophan, it is probably the best stress-buffering food that a pregnant woman can eat. Studies also suggest that probiotic consumption during pregnancy may confer health benefits to both mother and baby. In general, what all of these nutritional practices have in common is that they each rein in the exaggerated immune responses and inflammation resulting from chronic stress. In so doing, the simple act of taking supplements or changing what you eat in pregnancy (or in the case of vitamin D, getting enough sunlight) is suggested to protect against neurodevelopmental disorders arising from inflammatory brain damage to the fetus. As always, it's best to review any supplements with your doctor, particularly when pregnant. Unfortunately, many obstetricians do not formally recommend vitamin D supplementation, especially if they hear that a woman is taking a prenatal vitamin. However, the typical vitamin D doses in prenatal vitamin supplements often fall short of what pregnant women should ideally be getting (particularly if you are naturally darker-skinned or if you regularly apply UVB-blocking sunscreen, both of which block your body's ability to synthesize vitamin D from sun exposure).

When it is time for labor and delivery, additional strategies exist for an expectant mother to continue to swing the pendulum away from stress and distress and toward health. One of the best things a mother-to-be can do is to have a vaginal birth. When a baby is delivered vaginally, the baby's passage through mother's birth canal triggers an important relaxation nerve (the vagus nerve) in the mother. When this happens, the activated nerve causes oxytocin to start flooding the mother's brain. This oxytocin surge is evolutionarily smart not only to prime the mother for nursing and other socially attuned, maternal behavior, but also to lessen the pain of delivery given oxytocin's aforementioned ability to boost levels of natural painkillers. If a C-section is inevitable, immediate skin-to-skin contact has been shown to increase oxytocin and to provide health benefits to both mother and baby, paving the way for a health-promoting, in-tune relationship between mother and child. As a distinct but related aside, C-section mothers should also request the gauze technique, important for a baby's microbiome, which is discussed more in Chapter Seven.

PROMOTING EQUILIBRIUM

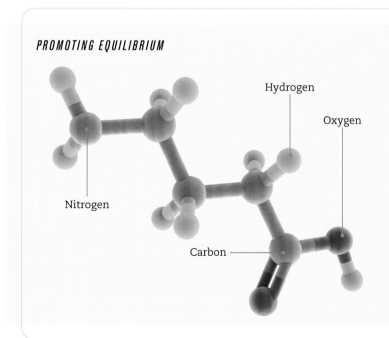

Hydrogen

Oxygen

Nitrogen

Carbon

GAMMA-AMINOBUTYRIC ACID (GABA) MOLECULAR MODEL

GABA is the main inhibitory neurotransmitter for the nervous system. It helps promote resting states and is important for homeostasis. Although some anti-anxiety medications work on GABA receptors, GABA levels increase naturally through meditation and yoga, and also from high-progesterone states such as pregnancy.

MODERN CHALLENGES
Balancing early life stress with family and work

In addition to the earliest months of existence in utero, the postnatal period is very important for our developing brain and stress response systems. It is during this critical early period that the dynamic between a mother and her infant translates into biological changes that may contribute to the child's temperament. Intense stress during this period can result in real physical changes that can persist into adulthood. These changes include alterations in important stress-regulating centers of our brain.

Perhaps the most significant source of early life stress is a discordant relationship or prolonged separation between mother and child. This is especially relevant with the rise of two-income households and the expectation for working mothers to return quickly to the workforce after having a baby. Given the critical importance of the early developmental years on the future mental and physical well-being of the child, it is really important for today's working parent to find a healthy balance between work and family and to foster an attuned relationship with his or her child.

Although the demands of balancing family and work can overwhelm both mother and father alike, mothers appear to be especially vulnerable to stress when it comes to caring for baby. This makes sense on an intuitive level, since a new mother is usually the one to wake up in the middle of the night to nurse her baby. In addition, her postpartum brain experiences drastically changing levels of hormones that will affect her reactivity to stress. Specifically, in the several weeks postpartum, the stress-lowering reproductive hormones mentioned earlier (estrogen and progesterone) cease to flood the mother's brain, increasing her general irritability levels and her vulnerability to stress. This heightened vulnerability to stress in turn increases her risk for depression, which can set off a number of events that negatively impact her and her baby's health.

A stressed and depressed mother is less likely to engage and interact with her baby. She is less likely to cuddle with, coo at, or play with her baby. By failing to provide sufficient and sensitive stimulation for her baby during important windows of brain development, the emotionally unavailable mother causes actual brain changes that can ultimately impair her child's ability to experience positive emotions in the future.

Extreme examples of discord between mother and child include child neglect and abuse. These increase a child's risk for developing psychiatric disorders later in life including mood and anxiety disorders, post-traumatic stress disorder, and antisocial behavior. Early genetic changes leading to dysfunctional stress regulation and defective oxytocin systems appear to mediate this increased risk.

Fortunately there are many strategies we can use to get our child's oxytocin system back on line and toward healthier stress responses. Studies show that something as basic as hearing a mother's voice or feeling her touch will boost oxytocin levels in her child following a stressful event. So if your child is having a bad day, talking to her or giving her a hug can raise her oxytocin levels and help her recover from stress more quickly.

One major stressor for babies is being born prematurely. Studies of premature infants show that frequent skin-to-skin contact, "kangaroo care," causes them to gain weight and reach important developmental milestones more quickly. They also recover from infections more quickly and cry less. As if that's not enough, mothers providing kangaroo care experience an increased sense of calm and well-being. Some practitioners believe that these benefits are rooted in the repeated "oxytocin doses" that result from all this physical contact.

Breastfeeding

Breastfeeding confers a wide range of health benefits to mother and child. What is remarkable is that the simple act of nursing your child can make you a more sensitive mother. Brain imaging studies of nursing mothers show that their brains react more sensitively to different infant cues. The increased sensitivity can make a mother more attuned to her child, increasing the likelihood that a baby's particular cries and vocalizations will be interpreted accurately. The end result is that baby's needs are more reliably met, helping baby to develop a secure attachment and preventing overactivation of stress systems that occurs when a distressed baby is not being sensitively cared for.

Finally, it appears that not all early-life stress is necessarily bad. Researchers in "stress inoculation" argue that some stress exposure is necessary for developing resilience or stress coping. So, you need to experience occasional stress to learn how to deal with it. The key is the duration and intensity of the stressor—if chronic, severe, or frequent enough to not allow for recovery, stress intoxication will cause you to feel burned out and will make you more vulnerable to becoming mentally and physically sick (see below).

STRESS DURING CHILDHOOD

Early life stress results in short- and longer-term effects that affect our genes, our brain structure, and brain function. Some early life stress is important to help a person learn adaptive skills and to foster resilience, but severe stress can lead to mental disorders and to health problems.

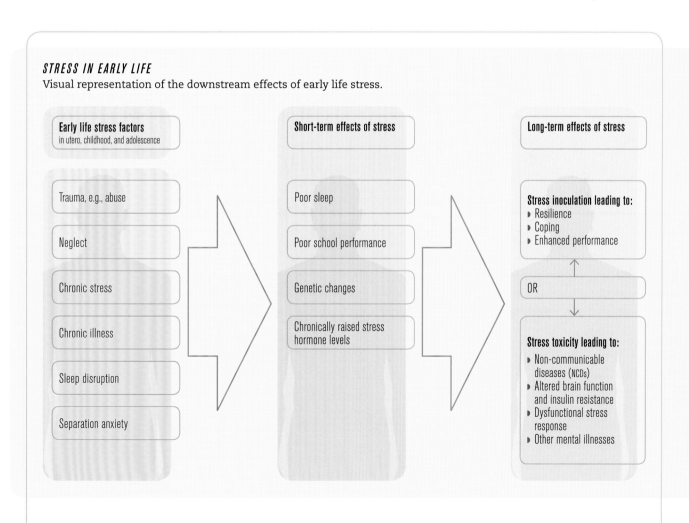

STRESS IN EARLY LIFE
Visual representation of the downstream effects of early life stress.

Early life stress factors
in utero, childhood, and adolescence

- Trauma, e.g., abuse
- Neglect
- Chronic stress
- Chronic illness
- Sleep disruption
- Separation anxiety

Short-term effects of stress

- Poor sleep
- Poor school performance
- Genetic changes
- Chronically raised stress hormone levels

Long-term effects of stress

Stress inoculation leading to:
- Resilience
- Coping
- Enhanced performance

OR

Stress toxicity leading to:
- Non-communicable diseases (NCDs)
- Altered brain function and insulin resistance
- Dysfunctional stress response
- Other mental illnesses

STRESS AND MENOPAUSE
How hormonal changes can lead to stress

Menopause is defined as the cessation of menses for twelve months, generally occurring in women between the ages of 49 and 52 years. The menopausal transition, "peri-menopause," often spans several years leading up to menopause and is heralded by changes in the menstrual cycle that are accompanied by wide fluctuations in reproductive hormone levels. Estrogen levels are erratic, and progesterone levels gradually decline, culminating in the eventual cessation of both ovarian hormones.

It is increasingly evident that these hormonal changes affect much more than a woman's reproductive abilities. Different levels of estrogen and progesterone can lead to different psychological states. Lower estrogen in menopause means lower serotonin, which in turn increases a woman's susceptibility to stress. Additionally, progesterone levels become depleted with menopause. This results in lowered GABA levels, further increasing her vulnerability to stress.

Mood changes in menopause

In addition to lower levels of these hormones, it is erratic hormonal fluctuations during menopause that wreak havoc on a woman's ability to deal with stress. This helps explain why mood changes are so common during this time. Thus, it is more than just the stress of this life stage symbolizing the end of a woman's reproductive years; it is the change in her hormone levels that directly affect her brain chemistry and stress reactivity. The connection between changing brain GABA levels, reproductive hormones, and stress axis dysfunction brings us to the concept of "reproductively based" mood disorders.

A woman with reproductive depression appears to have her depressed moods stem from altered levels of reproductive hormones. Her sensitivity to hormonal shifts leads to stress axis dysregulation and low mood. If you are a woman, one way to tell you are suffering from reproductive depression is to pay attention to the timing of your low moods. If they occur fairly predictably at monthly intervals, and if you also had a normal mood during pregnancy, but struggled with postpartum depression, you may be a woman who is more vulnerable to estrogen and progesterone shifts. Management would involve hormone-steadying strategies.

Notably, postmenopausal depression may be less responsive to hormone-balancing treatment since estrogen and progesterone fluctuations are no longer the issue. Rather, low but steady levels of estrogen are more characteristic of women who have gone past menopause, with a corresponding decrease in stress-buffering serotonin and oxytocin levels. Thus, postmenopausal depression may better respond to

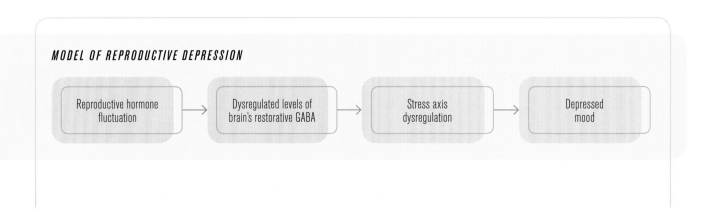

MODEL OF REPRODUCTIVE DEPRESSION

Reproductive hormone fluctuation → Dysregulated levels of brain's restorative GABA → Stress axis dysregulation → Depressed mood

more conventional serotonin-boosting treatment, and to strategies mentioned earlier that raise oxytocin levels, which improves feelings of well-being.

One other important menopause-related change has to do with cellular metabolic wear and tear. After menopause, women are more vulnerable to oxidative stress, in part because their lower progesterone levels make them less able to activate a progesterone-regulated anti-inflammatory gene, leaving them more stressed and more inflamed.

Decrease in melatonin

Finally, it appears that menopause leads to decreased levels of another crucial hormone: melatonin. Melatonin has many important functions: it is a powerful antioxidant, it promotes sleep, it establishes our circadian rhythm, and helps our brain cells grow. Although most melatonin is produced by our brain's pineal gland, ovaries and testes are another source of it. Decline in melatonin levels associated with the menopause may help explain why women have sleep difficulties as they get older, and why there is a higher prevalence of (oxidative) stress-related disease including cardiovascular disease and dementia following menopause.

Improving life with the menopause

Fortunately, there are things we can do to mitigate the effects of declining hormones that come with menopause. Low serotonin levels can be mitigated by eating tryptophan-rich foods and getting enough vitamin D. We can boost brain GABA levels by practicing meditation. We can help fight inflammation and oxidative stress by eating healthy, antioxidant-rich foods. We can restore melatonin levels by minimizing light exposure at night, including blue-light-emitting electronic devices.

The future of women's mental health will likely expand on the growing body of hormone-informed knowledge. Although all the details can feel overwhelming, common themes emerge. Namely, heightened stress reactivity, oxidative stress, and inflammation all increase our susceptibility to disease, whereas interventions that dampen chronic stress and inflammation increase our likelihood of

HEALTH BENEFITS OF DOG OWNERSHIP
Dog ownership appears to confer a wide variety of health benefits that may be especially relevant during menopause. Dogs encourage outdoor physical activity, which helps boost levels of endorphins and vitamin D. Petting a dog can also increase stress-buffering oxytocin and serotonin levels, which become depleted during the menopause. .

enjoying physical and mental health through all stages of life. The more healthily we eat, the more physically active we are, the more mindful and socially connected we are, even the more in tune with the sun (and darkness) we are, all these things generally translate into less oxidative wear and tear on our bodies and, as a result, improved well-being.

Chapter Seven

STRESS AND NUTRITION

Throughout this book, we see that chronic stress overwhelms our body's adaptive capabilities and increases our risk of developing disease. From a purely economic standpoint, stress is extremely costly to the body because it makes us spend a great deal of energy fighting threats to our safety—be they real, imagined, or exaggerated. To meet these increased energy demands, a constant supply of fuel (food) is needed. Otherwise, we will quickly be running on empty. This chapter examines the dynamic relationship between food and stress, and the complicated role nutrition plays in stress biology. Indeed, the notion that food simply serves to supply us with energy reserves to meet the challenges of everyday life is an outdated one, increasingly replaced by the idea that food directly impacts our physical and mental health and the way we respond to stress. On the one hand, we will see that non-nutritive foods typical of Westernized diets can actually serve as a stressor in and of themselves. On the other hand, we will see that nutritive foods rich in phytochemicals can be restorative, dampening the metabolic wear-and-tear effects of stress. To appreciate this dual role of food, we will examine how nutrition affects the interrelated concepts of oxidative stress, inflammation, and the microbiome or "gut–brain axis."

WHAT YOUR BRAIN NEEDS
Fueling the brain

To understand why nutrition is so important, we first need to explore how our brains work under normal, everyday circumstances. Although the adult human brain comprises only 2 to 3 percent of the body in terms of weight, it consumes an impressive 20 percent of the body's energy. This high metabolic activity takes place even during restful states, such as reading or sleeping. In order to meet its basic energy needs, the brain relies on fuel exclusively in the form of glucose—a monosaccharide (or simple sugar) derived from the different foods that we eat.

When we eat, digestive enzymes break food down into starches and sugars, the simplest of which is the glucose molecule your brain relies on. Insulin, an important hormone found throughout our bodies, then binds to insulin receptors on cell membranes to help deliver glucose into the cells. Once inside the cells, a complex process takes place that ultimately converts glucose into a usable form of chemical energy called adenosine triphosphate (ATP). This is an oxygen-dependent and enzyme-dependent process called "oxidative metabolism." Because of its efficiency, oxidative metabolism is the default mode by which all air-breathing animals generate energy. Its efficiency, however, comes at a cost in the form of "oxidative stress." If left unchecked, oxidative stress disrupts homeostasis, leading to cell damage ("oxidative injury") and cell death. Notably, excess energy production caused by high caloric intake is one of the ways to tip the balance and cause oxidative cellular injury. Not surprisingly, high-calorie diets have been linked to impaired cognitive function.

So far, we have seen we need glucose and oxygen to be converted into energy for proper brain function. However, we cannot directly dump these into our brains. Instead, these cellular-energy precursors are delivered by blood vessels, which are lined with endothelial cells. Endothelial cell integrity is important for normal cerebral blood flow and for the delivery of oxygen and nutrients to the brain. Thus, even with a well-rounded diet and good absorption of nutrients from your digestive tract, if your endothelial cells are impaired, your brain will suffer as a result of compromised delivery of oxygen and nutrients to the brain.

Nutrients

If we assume that the absorption and delivery of nutrients is not compromised, it is then important to talk about the dietary intake of nutrients. We need a variety of vitamins, minerals, fatty acids, and amino acids for proper cellular functioning and health. Dietary deficiencies can result in various forms of impaired neuronal functioning, including but not limited to the following:

▸ Improper energy utilization for example, B vitamins and magnesium serve as crucial cofactors for enzymes involved in energy harvesting;
▸ Neurotransmitter depletion for example, dietary tryptophan is an important precursor to the stress-reducing neurotransmitter serotonin; and calcium deficiency results in impaired neurotransmitter release;
▸ Impaired neurotransmitter receptor activity and immune function for example, as with Vitamin D deficiency;
▸ Impaired cell membrane fluidity for example, as with omega-3 fatty acid deficiencies;
▸ Decreased levels of brain-derived neurotrophic factor (BDNF)—a brain growth factor important for neuron survival, growth, and development.

The table opposite provides a list of essential nutrients thought to be important for overall health. As alluded to above, studies in nutritional neuroscience, endocrinology, and cardiology all increasingly point to the common theme that

Micronutrient	Functions (abbreviated list)	Effects on cognition and emotion	Common food sources
Vitamin A	Antioxidant Maintains normal vision and immune function	Antioxidant vitamins delay cognitive decline	Eggs, butter, liver, carrots, sweet potatoes, spinach, kale
B vitamins	Generation of cellular energy DNA repair Red blood cell production	Positive effects on memory performance, reduction in cognitive impairment	Meat, beans, fortified grains and cereals Vegans are at risk of B12 deficiency
Vitamin C	Antioxidant, dopamine production, iron absorption Supports immune function	Delay cognitive decline Mood enhancement	Kiwi, spinach, kale, oranges, tomatoes
Vitamin D	Supports immune function and bone health Nerve health and production of serotonin	Helps preserve cognition and improve mood	Fatty fish, milk, eggs UVB radiation from sunlight
Vitamin E	Antioxidant Supports cell membranes and normal nerve function	Antioxidants delay cognitive decline	Olive oil, nuts, spinach, avocados, asparagus, wheat germ
"Vitamin" F (omega-3)	Anti-inflammatory activity Supports cell membranes and nerve function	Amelioration of cognitive decline and improves mood and anxiety symptoms	Fish, flax seeds, krill oil, algae, walnuts
Choline	Precursor to acetylcholine (neurotransmitter involved in memory, cognition)	Preserves cognition	Egg yolks, milk, peanuts, chicken, beef, liver
Calcium	Release of cytokines and neurotransmitters from cells (important for their function)	Calcium homeostasis required for optimal mood, cognition	Milk, dairy, sardines, tofu, collard greens
Magnesium	Involved in many important cellular chemical reactions Supports nerve, muscle, and blood vessel function	Reduces anxiety symptoms	Seeds, nuts, spinach, beans, milk, brown rice, cocoa, bananas
Iron	Transport of oxygen in blood, synthesis of hemoglobin Neurotransmitter synthesis, immune function	Normalizes cognition and energy Deficiency: anemia, ADHD, depression	Meat, fish, beans, lentils, spinach, eggs, raisins
Zinc	Healthy immune function Dopamine production	Antidepressant properties, protection against cognitive decline	Oysters, beans, nuts, whole grains
Selenium	Antioxidant. Supports thyroid function	Preserves mood and cognition	Nuts, cereals, fish, meat, eggs
Copper	Energy production, iron utilization Neuroprotection and neurotransmitter synthesis	Preserves cognition	Oysters, crab, clams, sunflower seeds, kale
Potassium	Maintains fluid and electrolyte balance Supports proper nerve conduction, muscle contraction	Preserves cognition	Beans, potatoes, raisins, bananas, spinach, tomatoes
Iodine	Ensures proper energy metabolism of cells Maintains healthy thyroid function	Preserves cognition, mood enhancement	Iodized salt, fish, algae, some vegetables and dairy
Arginine	Nitric oxide (precursor for blood vessel dilation) Secretion of growth hormone	Preserves cognition	Meat, fish, dairy

what is healthy for the brain is also healthy for the immune, cardiovascular, and endocrinologic systems (and vice versa). The common thread appears to be the highly interrelated concepts of inflammation, oxidative stress, and immunoregulation. Indeed, dysregulated immunity, endothelial dysfunction, and inflammation are increasingly understood to be major drivers of chronic, non-communicable disease. Finally, the role genetic differences play in a person's response to different nutrients is being explored. There is increasing evidence that genetic differences affect how we metabolize drugs and nutrients.

FOOD AND PSYCHOSOCIAL STRESS
How stress depletes our nutrient intake

We have just reviewed what your brain needs under everyday circumstances. What happens to these needs during stress? To answer this question, we first need to review what happens to us when we are stressed.

As we have seen in previous chapters, stress activates the hypothalamic-pituitary-adrenal (HPA) axis and the sympathetic nervous system (the "fight-or-flight" branch of our nervous system), which results in an outpouring of the stress hormones cortisol and epinephrine (adrenaline). With epinephrine, our pulse rate goes up, and our heart pumps blood more rapidly to our extremities. Aldosterone (a hormone) also gets secreted at higher levels, to ensure our body holds onto necessary fluids, which raises our blood pressure. Cortisol works to keep blood glucose elevated to ensure that our cells are properly fueled while we are facing a threat. Large quantities of excitatory amino acids are released, helping our cells communicate more rapidly with each other (for faster speed of thought, necessary when we have to "think fast" in a dangerous situation). Nuclear factor NF-kB (see pages 62–3) is activated, which sets us into a pro-inflammatory state: endothelial cells lining our blood vessels release inflammatory cytokines, and immune cells become activated, resulting in the synthesis and release of more inflammatory cytokines, acute phase proteins, and free radicals. This stress system activation is highly coordinated and efficient, but it expends a great deal of energy. In response, we have important negative feedback systems to allow for the restoration of homeostasis. Once the threat to our safety subsides, cortisol signals the hypothalamus to stop releasing corticotrophin-releasing factor, which shuts off pituitary secretion of ACTH, which finally causes our adrenal glands to stop secreting cortisol.

Unrelenting stress and nutrition

Unfortunately, as we see throughout this book, modern day psychological stress tends to be unrelenting. Chronic stress results in constant metabolic wear and tear, which risks overwhelming our body's defenses and renders us vulnerable to disease. Moreover, the hypermetabolic state of chronic stress increases our nutrient demands, and depletes us of important vitamins, minerals, and antioxidants that are required to restore cellular physiologic stability. If we are not careful, we can become deficient in various nutrients in as little as a few weeks. Micronutrient deficiency in turn poses an additional stress on our system, thereby adding fuel to the fire.

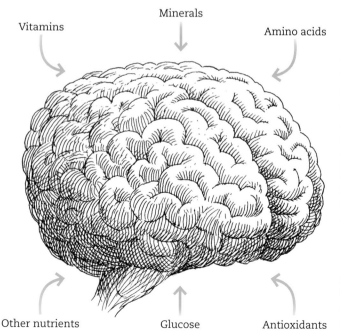

Vitamins
Minerals
Amino acids
Other nutrients
Glucose
Antioxidants

BRAIN FOOD
Without a varied diet rich in healthy foods, the brain would lack the nutrients it needs to function properly. Nutritive foods are especially important during stress since they help restore homeostasis.

MEDITERRANEAN DIET
Food pyramid of the typical Mediterranean diet that has been associated with longevity and health. A key aspect is the high amount of vegetables, beans, nuts, and healthy fats and fish.

Fortunately, as reviewed in other chapters, we have heterostatic (or set-point changing) tools to enhance our body's natural defenses and buffer against the metabolic wear-and-tear effects of stress. Nutrition is an important heterostatic tool. It aids the brain as our allostasis monitor to maintain stable physiologies in the face of changing external and internal environmental challenges. Like mind–body approaches, although not curative, it can have the benefit of improving our defenses and maintaining health.

Different dietary patterns have been studied that demonstrate health benefits and may have a role in fighting stress-related, non-communicable diseases such as the metabolic syndrome, cancer, depression, and dementia. Examples include the Mediterranean diet, the anti-inflammatory diet, the Paleolithic diet, and the low glycemic index diet. Each of these diets has the following feature in common: they echo diets our ancestors conceivably followed by emphasizing heavy intake of "functional foods"—fruits, vegetables, nuts, fish, and lean meats—while crucially minimizing refined grains and sugars typical of westernized diets (see above and the table on page 127).

Interestingly, although the Mediterranean diet is the most widely studied, research suggests that the Paleolithic diet, with its avoidance of grains (which are dense, acellular carbohydrates, unlike the cellular carbohydrates of fruit, leaves, and tubers) confers additional health benefits. Regardless, adherence to each of these functional-food based dietary patterns is thought to provide health benefits by restoring homeostasis, and decreasing oxidative stress and associated inflammation, thereby allowing for the repair of cellular physiologic machinery. We begin to feel reenergized and reinvigorated, ready for the next threat that comes along, be it real or imagined.

ENDOTHELIAL HEALTH

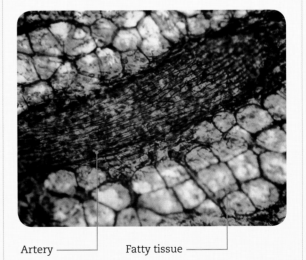

Artery —————— Fatty tissue ——————

AN ARTERY IN FAT TISSUE
A light micrograph of an artery (center) running through fatty tissue, showing endothelial cells lining the artery. Endothelial health is increasingly appreciated as a marker of general health, since endothelial injury or damage results in the release of harmful inflammatory cytokines that disrupt homeostasis. If inflammation goes unchecked, it leads to disease.

THE PROBLEM OF COMFORT FOOD
Stress and bad nutrition choices

People often make poor food choices. During the transition to a modern diet, the intake of refined grains, sugars, and saturated fats has drastically increased. This, in combination with increasingly sedentary lifestyles and the chronic psychological stress of modern times, explains why obesity and the metabolic syndrome are pandemic.

Comfort food is the term used to describe palatable, high-calorie, low-nutrient foods typical of westernized diets and are classically consumed in response to emotional stress. Typical foods that have added amounts of sugar, saturated fat, and/or sodium include pizza, ice cream, and white bread. "Processed foods" usually refer to foods that have been designed by the food industry to be particularly rewarding by adding fat and sugar in high concentrations, while at the same time stripping food of its water, fiber, and protein content in order to make it rapidly absorbable. Researchers have likened these particular characteristics of processed or comfort food to drugs of abuse such as cocaine.

Epidemiologic studies suggest that perceived stress motivates less healthy food choices. For example, it has been shown that female college students who reported more perceived stress ate more sweets and fast food, and less fruits and vegetables. Research suggests that this behavior may be a form of self-medication: recent studies in rodents and humans suggest that sucrose consumption can temporarily lower stress-induced increases in cortisol, thereby making some people under stress more hooked on sugar by "self medicating." Given how stressed people are, it should come as no surprise that the average American consumes approximately 79 pounds of added sugar per year. Interestingly, dopamine—a key

neurotransmitter involved in reward signaling, pleasure, and seeking behavior—has been shown to become dysregulated with both drug addiction and processed food consumption.

The glycemic index
Compared to naturally occurring foods, highly processed foods are more likely to induce a spike in blood sugar levels and insulin levels. The term used for this aspect of food is "glycemic index." A study shows that eating a high-glycemic index food results in activation of the nucleus accumbens, the brain's reward center, which is implicated in drug addiction.

So why is this so bad for us? Not only does the addictive nature of processed food make it extremely difficult to substitute these poor food choices for healthy alternatives, but also the high glycemic index of these foods is thought to trigger a whole cascade of pathologic events in our bodies. In fact, many of the same physiologic changes that happen with psychological stress also occur when we eat processed food. Thus, for the chronically stressed individual, processed comfort foods of the westernized diet compound the metabolic disarray that occurs with stress. Studies suggest that diets high in refined grains are linked to diabetes, obesity, heart disease, depression, and other chronic diseases.

After a high-glycemic meal, our body is flooded with glucose and triglycerides. This provokes spikes in pro-inflammatory mediators (IL-6 and CRP), leading ultimately to endothelial dysfunction and the formation of fat in arterial walls. Sugar has been shown to activate our old friend (or rather, foe) NF-kB, resulting in more pro-inflammatory cytokines and lower levels of brain-derived neurotrophic factor

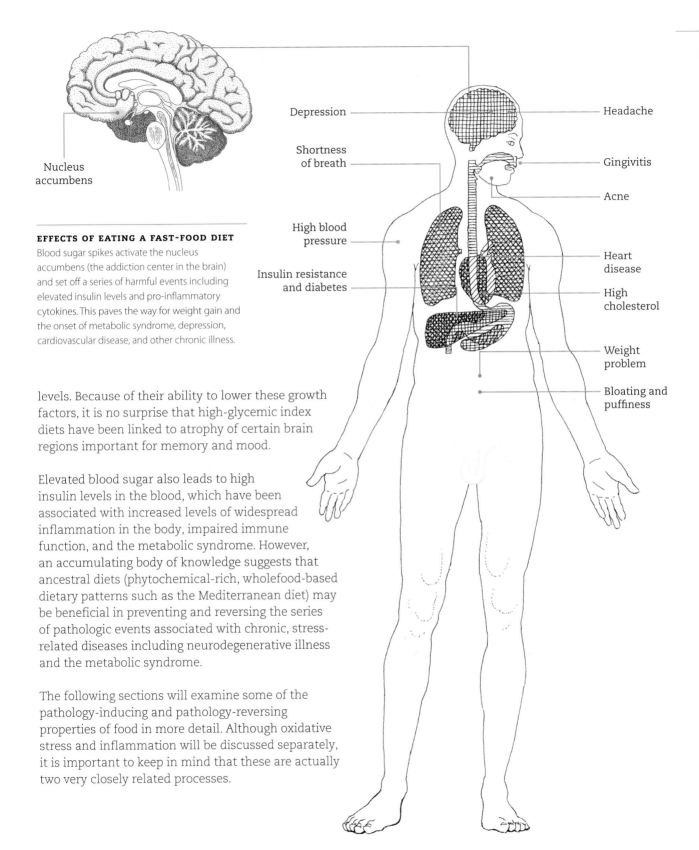

Nucleus
accumbens

Depression

Headache

Shortness
of breath

Gingivitis

Acne

High blood
pressure

Heart
disease

Insulin resistance
and diabetes

High
cholesterol

Weight
problem

Bloating and
puffiness

EFFECTS OF EATING A FAST-FOOD DIET
Blood sugar spikes activate the nucleus
accumbens (the addiction center in the brain)
and set off a series of harmful events including
elevated insulin levels and pro-inflammatory
cytokines. This paves the way for weight gain and
the onset of metabolic syndrome, depression,
cardiovascular disease, and other chronic illness.

levels. Because of their ability to lower these growth
factors, it is no surprise that high-glycemic index
diets have been linked to atrophy of certain brain
regions important for memory and mood.

Elevated blood sugar also leads to high
insulin levels in the blood, which have been
associated with increased levels of widespread
inflammation in the body, impaired immune
function, and the metabolic syndrome. However,
an accumulating body of knowledge suggests that
ancestral diets (phytochemical-rich, wholefood-based
dietary patterns such as the Mediterranean diet) may
be beneficial in preventing and reversing the series
of pathologic events associated with chronic, stress-
related diseases including neurodegenerative illness
and the metabolic syndrome.

The following sections will examine some of the
pathology-inducing and pathology-reversing
properties of food in more detail. Although oxidative
stress and inflammation will be discussed separately,
it is important to keep in mind that these are actually
two very closely related processes.

FOOD AND OXIDATIVE STRESS
Is our diet putting our bodies under stress?

As explained in previous chapters, "oxidative stress" refers to an imbalance between the production of reactive oxygen species (ROS) and our body's ability to neutralize these reactive intermediates or to repair the damage caused by them. The drivers of oxidative stress originally evolved as an important defense mechanism against invading pathogens (for example, viruses and bacteria) and continue to have that role. As with most things in nature, it is when the biological system becomes tipped too much in one direction that problems occur.

What causes oxidative stress?

Each time we consume oxygen and generate usable cellular energy, a chain of events occurs resulting in reactive oxygen species. These highly reactive intermediates lead to free radical formation, which in turn causes cellular damage. If the cellular damage exceeds the cell's natural repair abilities, as with aging and disease, programmed cell death (or apoptosis) can occur. Fortunately, Mother Nature has provided us with two natural defense systems against oxidative injury: an intrinsic, enzymatic antioxidant defense system as well as a non-enzymatic antioxidant defense system that includes endogenous antioxidants (Coenzyme Q10) and exogenous antioxidants, i.e. those ingested from foods.

Normally, there is a balance between oxidants and antioxidant defenses. With stress, aging, and processed food consumption, pro-oxidant forces outnumber antioxidant defenses, and oxidative stress results. Studies have implicated oxidative stress as a driver for glycation, an important form of protein damage relevant to medicine given its suggested role in disease pathogenesis. Glycation occurs when a reducing sugar such as glucose or fructose reacts with a protein. "Advanced glycation end products (AGEs)" is the term for these dysfunctional "sugarized" proteins. They have been implicated in a growing number of diseases, including Alzheimer's disease and cataracts.

When we eat, food needs to be broken down for us to be able to convert it into energy. This everyday process produces oxidants such as superoxide radicals and hydrogen peroxide. Refined grains are bad because the process of refining food eliminates much of the fiber, vitamins, minerals, and other nutrients that help protect us from tipping the balance toward oxidative stress. Additionally, refined grains and other high-glycemic index foods can rapidly increase sugar and insulin levels, and this generates free radicals. In a relaxed individual, this may not be a problem, since our intrinsic antioxidant enzymes can usually restore homeostasis. However, in the chronically stressed individual, regularly adding high-glycemic index foods to the mix overwhelms cellular repair mechanisms, paving the way for unchecked oxidative stress, biological aging, and disease.

ANTIOXIDANT POWERHOUSES

Blueberries are well known to be "antioxidant powerhouses" and as such have been studied for their health benefits. Nutritional neuroscience studies suggest that blueberry-supplemented diets reverse the age-related decline in hippocampal heat shock protein, improve object recognition memory, and decrease nuclear factor-kappa B ("Nuclear factor NF-kB" previously) levels in aged rats. Nut consumption has also been shown to improve cognitive and vascular function vis-à-vis improved lipid profiles and endothelial function, as well as lower levels of inflammatory markers, possibly more so when their skin is intact, as it contains polyphenols.

Antioxidants

Importantly, psychological stress, poor diet, and aging lead to the depletion of one of our most powerful endogenous antioxidants: glutathione. Researchers have demonstrated a correlation between glutathione levels and health; it was shown that healthy young people had the highest levels of glutathione, whereas levels were shown to be lower in healthy elderly and lower still in sick elderly people. Glutathione's antioxidant activity stems from the fact that it contains sulfur, which can neutralize free radicals, protect cells from damage, and help restore homeostasis. When glutathione levels are low, cellular repair mechanisms become overwhelmed, and stress-related diseases and aging could ensue. Fortunately, adding sulfur-rich foods (onion, garlic, leafy greens) to our diet can help restore glutathione levels. Moderate exercise and relaxation practices can also do this. However, even with normal glutathione levels, proper recycling and production of glutathione appears to rely on a variety of micronutrients that come from our diet, including B6, B12, folate as well as vitamins C and E. Dietary antioxidants include vitamins A, C, E, and plant polyphenols (including carotenoids and flavonoids, and tea catechins). These antioxidants work together with our endogenous antioxidant systems to help restore the cellular physiologic stability threatened by free radical damage, and may be able to stave off stress-related diseases and aging. Many clinical studies demonstrate health benefits of a wide range of dietary antioxidants.

It is important to note that supplementing with a particular nutrient in isolation may have inadvertent consequences. For instance, although the most widely studied and biologically active form of vitamin E is alpha tocopherol, it is suggested that other forms of this vitamin are important for health and can be depleted if alpha tocopherol is given in isolation.

By getting our antioxidants directly from fruits and vegetables as opposed to a pill, we have the added benefit of ingesting a natural balance of antioxidants as well as the associated fiber of the fruit or vegetable, which has its own metabolic benefits.

THE ANCESTRAL DIET'S EFFECT ON BRAIN FUNCTION
Ancestral and other wholefood-based diets are rich in nutrients that support brain function through a variety of mechanisms. These include helping your brain cells last longer, using energy better, communicating better, and making more feel-good, stress-buffering brain chemicals.

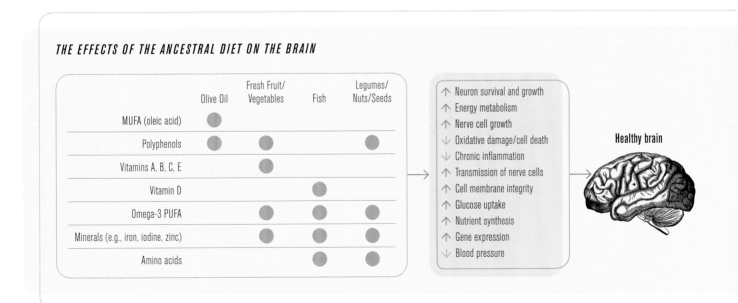

THE EFFECTS OF THE ANCESTRAL DIET ON THE BRAIN

	Olive Oil	Fresh Fruit/Vegetables	Fish	Legumes/Nuts/Seeds
MUFA (oleic acid)	●			
Polyphenols	●	●		●
Vitamins A, B, C, E		●		
Vitamin D			●	
Omega-3 PUFA		●	●	●
Minerals (e.g., iron, iodine, zinc)		●	●	●
Amino acids			●	●

↑ Neuron survival and growth
↑ Energy metabolism
↑ Nerve cell growth
↓ Oxidative damage/cell death
↓ Chronic inflammation
↑ Transmission of nerve cells
↑ Cell membrane integrity
↑ Glucose uptake
↑ Nutrient synthesis
↑ Gene expression
↓ Blood pressure

Healthy brain

FOOD AND INFLAMMATION
How our diet can reduce stress

Inflammation, as we have learned in other chapters, is the immune system's response to injury or infection. When it is time-limited and localized, it is important for wound healing. However, psychological stress, in the absence of bodily injury or infection, activates widespread inflammation since it prepares the body for possible injury in the face of perceived danger. Activation of the HPA axis activates NF-kB, which boosts pro-inflammatory cytokine production. Chronic stress, then, results in chronic, low-grade, systemic inflammation. It is this variety of inflammation that is a common thread in the vast majority of non-communicable, stress-related diseases. Relaxation practices, social connectedness, and sleep, as we have seen, can each inhibit pathogenic pro-inflammatory pathways.

Importantly, food can also modulate these same key pathways, for bad (pro-inflammatory) or for good (anti-inflammatory). Generally, foods that increase oxidative stress are also pro-inflammatory. On the other hand, antioxidant-rich foods are generally thought to be anti-inflammatory. As we will see, it is not just the quality but also the quantity of the food we eat that can influence the inflammatory response.

A BALANCED DIET

The benefits of a balanced diet have been known for centuries. Balanced diets rich in nutritive foods have inflammation-fighting properties that can offset the widespread, unchecked inflammation associated with chronic stress. *Still life with Seafood* painted by Jacob Fopsen van Es in the seventeenth century.

Pro-inflammatory dietary patterns

First, it is important to recognize the connection between pro-inflammatory dietary patterns and disease. Going back to the concept of advanced glycation end products (AGEs), it appears that the way food is prepared can affect the amount of AGEs in our food and in our blood. Animal products that have been fried, grilled, or broiled are higher in AGEs than animal products that have been boiled or stewed. However, the nature of the food itself is also important. Vegetables, regardless of cooking method, are lower in dietary AGEs. Different sugars appear to be more likely to lead to AGEs than others. For instance, fructose is considered to be more reactive than glucose, which makes it a bigger source of problematic AGEs. Studies suggest that the more AGEs in your diet, the more AGEs there will be in your blood, and the higher levels of biomarkers of inflammation such as CRP. Interestingly, an animal study of dietary AGEs demonstrated that rodents fed AGE-rich diets had shorter life spans than those rodents fed AGE-poor diets, even though the caloric intake and fat content of both groups was equivalent.

Another dietary culprit for increased inflammation is omega-6 fatty acids. They are derived primarily from refined vegetable oils (for example, corn, sunflower, and safflower) and have been shown to increase the production of pro-inflammatory cytokines. Westernized diets are notoriously high in omega-6 fatty acids as opposed to the anti-inflammatory omega-3 fatty acids. Dietary sources of omega-3 fatty acids include fish, walnuts, and wheat germ. Studies show that lower omega-6-to-omega-3 ratios are associated with lower pro-inflammatory cytokine production. Randomized, placebo-controlled studies demonstrate that blood levels of omega-3 fatty acids are inversely related to pro-inflammatory cytokines and depressive symptoms. Omega-3 fatty acids have been shown to diminish NF-kB activation while modulating stress-induced changes in the HPA axis.

Anti-inflammatory dietary patterns

Given the link between inflammation and mood disorders, interventional studies have investigated the influence of anti-inflammatory dietary patterns on depressive symptoms. The Mediterranean diet has

CURCUMIN

Curcumin, the phytochemical that gives the Indian spice turmeric its vibrant yellow color, has well-established anti-inflammatory properties and has been shown in human studies to have an antidepressant effect. Interestingly, it is also known to have an insulin sensitizing and anti-hyperglycemic effect, suggesting it may have potential in the management of the metabolic syndrome.

been shown to lower the risk of developing depression, presumably due to its anti-inflammatory effects. One large study demonstrated that a traditional dietary pattern of vegetables, fruit, meat, fish, and whole grains was associated with lower odds of developing depression and anxiety than a westernized dietary pattern high in refined, processed foods; however, causality could not be established.

Caloric restriction is another dietary way to reduce inflammation and possibly promote longevity. Daytime fasting, in the absence of any weight change, has been associated with decreased IL-6 and CRP levels compared with non-fasters and compared with pre-fasting levels. Caloric restriction is thought to promote cell survival by activating specific "sirtuin" anti-aging genes. Certain highly popularized "fountain of youth" antioxidants such as resveratrol and nicotinamide riboside also claim to promote longevity and protect against aging because of their ability to activate sirtuin genes and decrease inflammation.

FOOD AND THE MICROBIOME
How stress affects valuable gut bacteria

We have seen how our food choices and stress levels can impact our health by affecting our macro- and micronutrient status and by increasing or decreasing oxidative stress and inflammation. A more indirect yet powerful mechanism involves the way our diet and stress impact the gut microbiome and the gut–brain axis.

The microbiome
The microbiome refers to both benign and pathogenic microorganisms that share our body space, whether it is in our gut or on our skin. The human gut is colonized by a staggering number of microbes, over ten times more than the total number of human cells. Bacteria are an important component of the microbiota ecosystem in the human gut and comprise 1,000 species approximately. The microbiome has been referred to as a "second brain" due to the well-established bi-directional communication between the microbiota, the gut immune system, and the brain. Its importance is perhaps reflected in the recently discovered fact that gut colonization by bacteria develops even before we are born, with the maternal microbiome influencing fetal gut colonization by way of the placenta.

Gut bacteria
Gut bacteria play important immunologic, metabolic, and perhaps neurobehavioral roles in human health. On the one hand, enhancement of the gut microbiome with certain dietary patterns or pre- or probiotic supplementation appears to affect the brain and behavior. On the other hand, stress and altered microbiota may increase gut permeability, allowing bacteria and bacterial antigens to enter the bloodstream and activate an inflammatory immune response (thus, we see that the concept of the microbiome offers one more explanation for stress-induced systemic inflammation).

Gut bacteria help the gut maintain normal, everyday functioning and enhance health. Everyday functions include regulating gut motility, aiding in digestion, producing vitamins, absorbing minerals, activating and destroying toxins. Gut bacteria are also essential for the metabolism of lignans and isoflavones (natural compounds present in a variety of foods) into their bioactive forms, which may protect against cardiovascular disease, osteoporosis, and certain types of cancer. Gut microbiota have also been shown to generate short-chain fatty acids by fermenting soluble fiber in the gut, which may confer metabolic benefits via gut–brain neural circuits.

Generally, research suggests that the more diverse the gut microbiota, the better for your health. Decreased diversity of the gut bacteria has been implicated in many chronic diseases, including inflammatory bowel disease, obesity, cancer, and autism. Although the precise pathogenic mechanism has yet to be established, it may involve defective immunoregulation which occurs with decreased microbiome diversity. Altered neurotransmitter signaling may also play a role.

Nutritional diversity
The good news is that by changing the foods we eat, we can increase the diversity and affect the activity of our microbiome, which may then confer physical and mental health benefits. Ancestral, functional food-based dietary patterns have been associated with greater gut microbiota richness and lower inflammatory markers. Probiotic supplementation is another way to increase microbiome diversity and has been shown to reduce depression, anxiety, and other stress-related behaviors. Data suggest that changes in gut microbiota can directly regulate pro-inflammatory genes and affect neurotransmitter synthesis. Indeed, it appears that certain strains of bacteria are particularly good at producing GABA

levels which may have stress and anxiety-reducing benefits. Not surprisingly, probiotic interventions are increasingly being studied for the prevention and management of depression, anxiety, and the metabolic syndrome.

Interestingly, the epidemic rise of allergies and autoimmune disease is thought to be due not only to the heightened stress levels in modern society, but also to the decreased diversity of our microbiomes, as we have become relatively more removed from nature (and many bacteria, fungi, and yeasts). Additionally, over-sanitizing practices as well as monotonous diets high in refined foods have led to decreased microbiome diversity. As a result, our gut immune system is exposed to fewer antigens, which breed immunoreactivity should even newer "benign" antigens come along.

In contrast, a highly diverse microbiome, with a rich array of antigens presented to our gut immune system, is thought to foster immunotolerance and protection from illnesses stemming from defective immunoregulation. Because of this, researchers hypothesize that living within one to two miles of a forest (with greater microbiota diversity in the air) confers improved immunoregulation and stress resilience. Related to this, exposure to dogs (and the germs they carry) can also increase our microbiome diversity and confer health benefits such as decreased incidence of asthma, allergies, and depression.

THE GUT–BRAIN–MICROBIOME AXIS

Altered gut physiology leads to altered brain physiology, and vice versa. Moreover, your health can be influenced by the predominance of bacteria in your gut. This is the general concept behind fecal transplantation as a possible treatment for both gastrointestinal and neuropsychiatric illness.

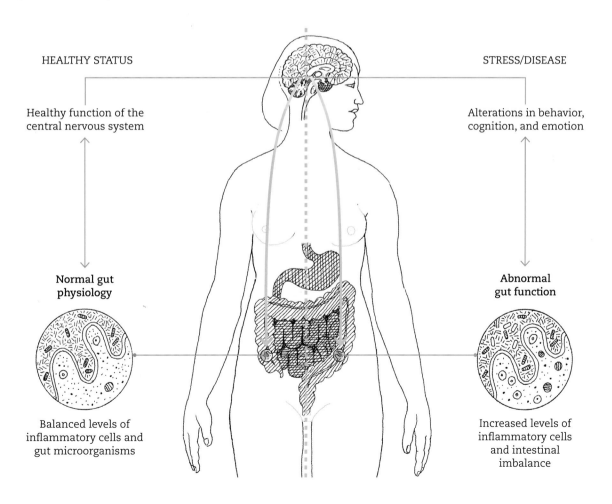

HEALTHY STATUS   STRESS/DISEASE

Healthy function of the central nervous system

Alterations in behavior, cognition, and emotion

Normal gut physiology

Abnormal gut function

Balanced levels of inflammatory cells and gut microorganisms

Increased levels of inflammatory cells and intestinal imbalance

Chapter Eight

STRESS, HEALTH, AND THE SOCIAL EXPERIENCE

While some organisms are solitary creatures, choosing to fend only for themselves and their offspring, humans are incredibly social creatures. From ancient civilizations to modern-day society, humans have combined their energy and resources to remain at the top of the food chain. Social relationships can benefit humans by offering care and support, cultivating feelings of social inclusion, and leading to reproductive success. The human brain is hard-wired for social interactions with a large neocortex, which is necessary for language, behavior and emotion regulation, conscious thought, theory of mind, and empathy. All of these skills allow humans to engage in high-level social cognition and interaction. Humans also rely on social interaction to maintain their physical and psychological health. In essence, social interaction is at the core of the human experience. In this chapter, we discuss the basic patterns of social interaction, how stress can arise from such interactions, the general consequences of prolonged stress, and how such stress can affect children. We also consider stress that arises from poverty or belonging to a minority group, and the physical and psychological illnesses that can result from stress.

SOCIAL INTERACTION
What is social interaction?

Social interaction consists of any behavior between two or more people that is geared toward affecting or acknowledging each other's thoughts, feelings, experiences, or intentions. For a behavior to qualify as a social interaction, those involved must be aware of each other and have the other people in mind while performing that behavior. People do not need to be in the presence of, or directly affect the other parties, to be socially interacting with them. For example, there is social interaction when people correspond by email, or when enemies ignore each other.

Five basic patterns of social interaction

There are five basic patterns of social interaction: exchange, cooperation, competition, conflict, and coercion. These patterns of social interaction are neither distinct nor mutually exclusive, and interact to form complex relationships between individuals, groups, and societies. What further complicates social interactions is that the quality of an interaction is subjective; what some people may perceive as cooperation, others may perceive as competition.

Social exchange

The concept of social exchange is based on the theory that people generally try to maximize rewards and minimize costs for themselves. In order to benefit the

COOPERATION

Cooperation is the social interaction in which people work together to reach common goals. Different parties can also cooperate to reach individual goals. These rowers need to cooperate with each other in order to row efficiently and to successfully compete against other teams.

most people and to lower costs, people engage in mutually beneficial social exchanges in which they try to help and avoid harming those who have helped them. Thus, a norm of reciprocity is formed where people expect beneficial exchanges to be repaid, such as recognition, friendliness, or gifts. The norm of reciprocity also results in the expectation that harmful exchanges will be reciprocated. As people are constantly trying to achieve the most advantageous benefit-cost ratio in their interactions with others, they will put the most energy into the relationships that they perceive to offer the greatest benefits at the lowest costs at that time, such as those with family members, business partners, and friends.

Cooperation

Cooperation is interaction in which people work together to reach common goals. While goals may not be identical between parties, as long as they are consistent in how they are achieved, different parties can cooperate to mutually reach individual goals. For example, in a teaching relationship, both the teacher and the student are working toward the student's proficiency in a certain subject. However, a student may be motivated to work hard at the subject to attain good grades, while the teacher may be motivated to aid in the student's success in order to maintain the reputation of being a good teacher.

Competition

Competition is similar to cooperation in that people strive toward common goals. However, rather than working together to achieve those goals, individuals or groups contend against each other in order to attain those goals for themselves. The presence of competition in society reflects that resources are limited, and therefore must be fought for. Resources could be physical, such as money, land, and water; or abstract, such as political support and attention.

Conflict

As with cooperation and competition, conflict involves a shared resource or goal. However, in conflicts, opposing individuals or groups seek to harm or get rid of other contestants. In a competition, competitors focus on reaching a goal rather than focusing on actions against other competitors: there

COERCION

Coercion involves social interactions in which an individual or group in power dominates another individual or group to achieve a certain goal. It usually involves use of force or intimidation to obtain compliance. In the painting *The Wolf and the Lamb* (c. 1819–20) by William Mulready, the artist has clearly and strikingly depicted that children too can be menacing and bully others.

are rules in place that restrict the extent of injury that can be dispensed by any party. But, in a conflict, those restrictions are ignored, and contestants will do whatever it takes to get what they want.

Coercion

Coercion involves social interactions in which an individual or group in power dominates another individual or group in order to achieve a certain goal. For example, a parent, with power over a child, demonstrates coercion when withdrawing his or her affection until the child complies and demonstrates a desired behavior.

STRESS AND CHILDHOOD
Can stress affect a child's brain in the early years?

Early childhood spans from conception to around five years old. In these early stages of life, critical physical, cognitive, and social-emotional growth and development occur. During early childhood, an individual's sense of self, attachment to others, and safety begin to develop. Early childhood is also the stage in which an individual's threat appraisal and response system is developing and easily modified by the environment. Because of this, children in this stage of life are especially vulnerable to the effects of psychosocial stress.

Psychosocial stress

Psychosocial stress is stress that arises from anything that disrupts or threatens to disrupt an individual's relationship with others. Having a low social status, being bullied, and meeting new people are all situations in which psychosocial stress can arise. Instances of social exchange, cooperation, competition, conflict, and coercion are often unpredictable, uncomfortable, or disturbing to an individual. In fact, psychosocial stress is the most common type of stress that individuals deal with.

How does the body respond to stress?

We have already learned that the body responds to stress by producing a fight-or-flight response, which prepares an individual to confront or avoid a real or perceived threat. When an individual is faced with a threat, the HPA axis is activated. The amygdala detects a threat and triggers the hypothalamus to release corticotropin-releasing hormone. This stimulates the pituitary gland to secrete adrenocorticotropic hormone, which promotes the increased production and release of the hormone cortisol. Increased levels of cortisol raise glucose production and blood pressure, and suppress the immune system. These reactions increase an individual's chances of success against a threat by boosting energy sources and decreasing the amount of energy spent on processes that are nonessential to managing the threat.

The risks of chronic stress

Although the stress response has been essential to human survival, and can be beneficial in some situations, it can also be harmful to the health of an individual. When stress is chronic, the HPA axis remains activated, leading to prolonged elevation of cortisol. Prolonged high levels of cortisol contribute to fat build-up, weight gain, suppressed thyroid function, an impaired cognitive performance, and a decrease in bone density and muscle tissue.

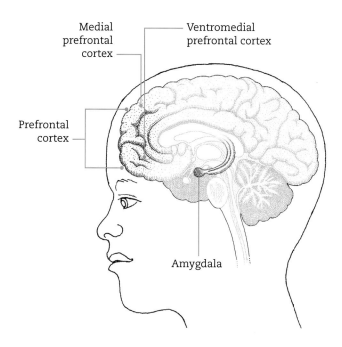

Medial prefrontal cortex

Ventromedial prefrontal cortex

Prefrontal cortex

Amygdala

THE EFFECTS OF STRESS ON A CHILD'S BRAIN
The prefrontal cortex (PFC) is involved in working memory and self-regulatory and goal-directed behaviors. Stress experienced in early childhood can impact the development of the prefrontal cortex, though such effects are not necessarily permanent.

Chronically increased levels of cortisol also weaken the immune system, rendering individuals more vulnerable to infection. High levels of epinephrine can also increase blood pressure over long periods of time, and lead to heart attacks or strokes. Not only that, chronically elevated levels of cortisol put an individual at a higher risk of developing stress-related disorders. Chronic stress can also result in hypersensitivity of the stress response system, making the system react more frequently, in greater magnitude, and take longer to return to baseline.

Effects of chronic stress in early childhood

Chronic stress experienced in early childhood can produce long-lasting detrimental effects. Prolonged elevation of cortisol levels puts children at risk of the many harmful effects mentioned above. There is also evidence that stress experienced in early childhood can impact the development of the prefrontal cortex. If a young child is subjected to a stressful event that he or she cannot overcome, that child is at a higher risk of reacting to future stressors with helplessness, and of developing attention regulatory problems. Chronic stress can also affect the development of the stress response system by making it more reactive to

any threat, and by making an individual assess more situations to be threatening. This perpetuates a vicious cycle throughout an individual's life in which that person is more vulnerable to becoming stressed, and as a result is chronically stressed.

Reversing the adverse effects of chronic stress

As mentioned before, threat appraisal and response systems develop during early childhood. While this does not bode well for a young child who is subjected to chronic stress, there are also some positive implications. While the systems are developing, they are easily malleable, or "plastic." Therefore, if a young child who was subjected to chronic stress is put into an environment with a consistently sensitive and responsive guardian, his or her threat appraisal and response systems could be reset, so that the child would react to stress appropriately and no longer suffer from the consequences of having an exaggerated stress response.

CHILDHOOD AND TOXIC STRESS
Severe stress in children

In the event of a threatening situation, people can have different types of stress responses: positive, tolerable, and toxic. These three categories, which produce different health outcomes, do not describe the threatening experience or stressor itself, but rather the effect of the stress response on the body.

Positive stress response

A positive stress response is indicated by an increase in heart rate and a slight elevation of hormone levels, both of which are short in duration. By reacting to new situations or threats with a positive stress response, an individual is prepared to react to a threat if necessary, but does not suffer from any long-term consequences of stress.

Tolerable stress response

A tolerable stress response is defined by a moderate and time-limited activation of the body's alert systems. This type of stress response lasts longer than a positive stress response, and is usually in reaction to threats that are more severe and difficult to handle. As long as the stress response does not last for too long, and the individual has social support in the form of caring relationships, the body will recover from the effects of a tolerable stress response.

Toxic stress response

A toxic stress response is characterized by an abnormally strong, frequent, and/or prolonged physiological reaction to a stressor. Such a response typically develops in reaction to extreme threats or traumatic experiences, such as physical or emotional abuse or neglect, exposure to violence, economic hardship, and substance abuse, or mental illness in a relative or caregiver. Toxic stress can have numerous harmful effects on an individual's health, like those mentioned for chronic stress.

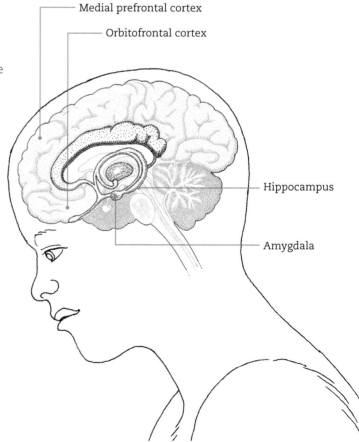

Medial prefrontal cortex

Orbitofrontal cortex

Hippocampus

Amygdala

A CHILD'S BRAIN AFFECTED BY DEPRESSION
Children can suffer from depression and when this condition is severe it can affect the developing brain and have long-term consequences that may continue into adulthood. This type of chronic stress is associated with hypertrophy (the enlargement of existing cells) and overactivity in the amygdala and orbitofrontal cortex, and loss of neurons and neural connections in the hippocampus and medial prefrontal cortex.

PREVENTION AND TREATMENT OF TOXIC STRESS IN CHILDREN

The sooner children who are suffering from adverse childhood experiences and toxic stress are identified, the better their prognosis. When children appear stressed, pediatricians, health workers, and teachers should be aware of risk factors such as social isolation, poverty, caretakers who are unemployed, low-educational achievement, single-parent homes, a non-biologically related male living in the home, family or intimate partner violence, young parental age, and caretakers demonstrating low self-esteem, substance abuse, and depression. Development of caring and supportive relationships with adults as early as possible is also crucial to preventing and treating toxic stress in children. Studies have shown that children who receive abundant parental warmth and affection, especially maternal warmth, have lower chances of developing health consequences in reaction to toxic stress. Other factors that help a child cope include having a structured school environment, involvement in an extracurricular activity, access to health care and social services, and therapy. Mind–body and mindfulness-based interventions, such as breathing techniques, guided imagery, meditation, and biofeedback, have also been shown to be useful.

THE EFFECTS OF STRESS ON A CHILD'S BRAIN
Learning in a supportive environment can be a positive stress for young children. Positive stress is necessary and promotes the ability in children to function competently under threat as they develop.

Adverse experiences and toxic stress

Adverse childhood experiences (ACEs) are distressing events that occur between conception and 18 years and have long-lasting physical and mental consequences due to the toxic stress they produce. Such experiences include witnessing or experiencing abuse, neglect, household dysfunction, domestic violence, crime, parental incarceration, parental discord, parental death, instances of racism or discrimination, substance abuse, economic hardship, and/or living with someone with a mental illness. Studies have shown that such experiences are quite prevalent, with 48 percent of children in the US, ages 0–17, experiencing at least one adverse childhood experience, and 23 percent experiencing two or more.

If a young child that has stable, supportive relationships with adults experiences a stressful situation, the care and guidance of those adults will act as buffers and bring the child's stress response system back to baseline. However, without stable relationships, children experiencing such toxic stress will develop an unbalanced stress response system. This can result in a multitude of persistent health problems that can appear during childhood and adulthood. During childhood, these adverse experiences put children at risk of permanently damaging their brain, altering their gene expression, developing a stress-related disease, getting infections more frequently and of greater severity, failing to develop effective stress-management skills, and developing mental illnesses such as depressive disorders, behavioral dysregulation, post-traumatic stress disorder, and psychosis. Adults with these childhood experiences are more prone to alcoholism and other substance abuse problems, poor coping skills, unhealthy lifestyle choices, lung diseases, obesity, and other chronic illnesses.

STRESS AND POVERTY
How living in poverty leads to stress

Globally in 2015, 836 million people live in extreme poverty, on under $1.25 a day, and over 160 million children under the age of five are short for their age due to a lack of food. One in seven children worldwide are underweight. In 2013, 14.5 percent of all Americans and 19.9 percent of American children lived in poverty. The 2015 federal poverty level guidelines established by the US government consider one person households making less than $11,770 a year, or four person households under $24,250 a year, to be living in poverty. In 2011, 24 percent of the general population and 27 percent of children in the European Union were at risk of falling into poverty and experiencing social marginalization. At the end of 2012, official data showed that China had 98.99 million people living below the national poverty line of RMB 2,300 per year. In 2011 in South Asia, 24.5 percent of its population lived in extreme poverty, on under $1.25 a day.

Poverty can be divided into two categories: absolute and relative. Absolute poverty is when a person lacks the basic necessities of survival such as food, shelter, and clean water. Relative poverty is when a person's standard of living and income is far worse than average in the area that they live in. In any case, living in poverty is stressful and can greatly impact someone's life.

The socioeconomic status-to-health gradient
There is a steady socioeconomic status (SES) to health gradient that exists in which the lower in socioeconomic status an individual is, the higher chance they will have conditions related to poor health. Those conditions include hypertension, heart disease, cancer, shorter life expectancy, depression, and schizophrenia. In some cases, people with the lowest socioeconomic status have a ten-fold increase in the chance of developing such conditions when compared to people with the highest socioeconomic status. Impoverished children and adolescents have an increased risk of being injured, and developing asthma, elevated blood pressure, and respiratory illnesses. Another study showed an inverse relationship between socioeconomic status and mortality, where those with a lower socioeconomic status die at higher rates than those with a higher socioeconomic status. A higher incidence of smoking, drinking, obesity, poor diet, a sedentary lifestyle, and a lack of access to health care in individuals with a lower socioeconomic status does not account for the majority of the socioeconomic status–health gradient. More and more research is now pointing to stress as being one of the primary factors that affects the health of those living in poverty.

The effects of poverty and stress
Poverty and stress are closely related. Living without adequate resources is inherently stressful. Stress arises not only from a lack of resources, but also from the chaos that often surrounds those who lack resources. From violence and residential mobility, to discrimination and unsanitary environments, the context of poverty increases the number of stressors that people have to deal with, and thereby reduces an individual's ability to deal with new threats.

Stress that stems from a lack of resources and the context that surrounds poverty is called poverty-related stress (PRS), which tends to be chronic. As with other cases of chronic stress, this type of stress can unbalance the body's stress response system and make an individual more vulnerable to becoming stressed, and developing some of many conditions correlated with chronic stress mentioned earlier. Stress stemming from economic hardships and strains has been found to lead to conditions such as depression, anxiety, and alcohol use.

TREATING POVERTY-RELATED STRESS IN LOW-INCOME FAMILIES

In treating poverty-related stress in low-income families, there are two important elements to target: the coping and relationship skills of the parents, and the coping skills of the children.

Research has shown that teaching skills to low-income parents with stress-related problems, about how to maintain healthy relationships, resolve conflicts, cope with stress, and practice child-centered parenting, reduces financial stress and improves coping and problem-solving skills. It has also been found that low-income children who are taught about the effects of stress on their health, emotional well-being, and coping skills, show decreases in involuntary engagement responses, such as intrusive thoughts

and rumination, and in directing problems toward themselves (internalization) and, thus, creating distress; or in expressing problems externally and thus generating conflict in the surrounding environment.

By using interventions to teach both parents and children coping strategies to deal with stress, low-income families can actively reduce the harmful effects of stress on their mental and physical health.

POVERTY IN THE USA

Poverty is associated with a lack of resources, a chaotic living environment, violence and discrimination; all these elements contribute to significant life stress.

MINORITY STATUS AND HEALTH
Do minority groups suffer more stress?

Globally, people who are native-born, male, able-bodied, heterosexual, cisgender, middle or upper class, educated, young adult-middle-aged, belong to the racial or religious group in power, and in good mental health, can consider themselves to belong to at least one majority group. In sociology, "majority" does not necessarily indicate a higher percentage of people relative to other groups, but a higher position of power or privilege in society. For example, although males generally only make up about half of the US population, in 2015, 80.7 percent of the House of Representatives, and 80 percent of the Senate were male. As men are often disproportionately in positions of power, they are considered to be a majority group while women are considered to be a minority group. In the European Union, more than 80 percent of respondents in each member state said there were times when lesbian, gay, bisexual, and transgender (LGBT) youth were bullied in school or received negative comments. In East and Southeast Asian cultures it has been documented there is a preference for fair skin as opposed to darker skin.

THE MINORITY STRESS THEORY
The minority stress theory proposes that societal challenges lead to added stress for minorities, which causes poorer health outcomes than for those not living in a minority environment.

Minority health disparities

Studies over the past thirty years have revealed mental and physical health disparities between majority and minority groups. In the US, non-Hispanic black adults are at least 50 percent more likely to prematurely die of heart disease or stroke than non-Hispanic white adults. They also have higher rates of colorectal, pancreatic, and stomach cancers, as well as depressive symptoms and substance abuse. Non-Hispanic blacks have an infant mortality rate that is more than double the rate for non-Hispanic whites. Prevalence of adult diabetes is higher in Hispanics, non-Hispanic blacks, and mixed race individuals than in non-Hispanic whites. One educational health disparity that exists is that adult diabetes is more common among adults without a college degree than those with one. Being of a lower socioeconomic status also has health consequences that have been mentioned in the section on the effects of poverty.

In terms of health disparities between genders, women are more likely to suffer from depression, anxiety, and somatic complaints. Additionally, lesbian, gay, bisexual, and transgender individuals show higher rates of suicide, substance abuse, depression, and anxiety throughout life. LGBT individuals are also at a higher risk of developing certain types of cancer and immune dysfunction.

Minority stress theory

The minority stress theory specifies that societal challenges lead to added stress for minorities, and that stress, as opposed to the challenges themselves, is the mechanism that causes poorer health. The theory also differentiates between external and internal stress processes. External sources of stress include rejection, prejudice, and discrimination. For example, LGBT individuals experience varying degrees of discrimination, or external stress, in different parts of the world. In general, acceptance is higher in North America, the European Union, and most of Latin America, while in most Muslim nations, Africa, parts of Asia, and Russia, homosexuality is more discriminated against. However, national surveys in the US have reported that 90 percent of LGBT youth have experienced prejudice in school.

Sixty percent of African Americans have experienced social rejection, employment discrimination, or housing discrimination; this reveals that there are many external sources of stress that come with belonging to a minority group. Internal stress processes, such as anxiety about discrimination and prejudice, memories of past negative experiences of bigotry, and feelings of inferiority and isolation, are considered to be proximal (internal) stress processes. For example, it has been found that LGBT individuals often display internalized homophobia, a reflection of societal discrimination, leading to self-hatred and low self-esteem. In combination, distal (external) and proximal stressors lead to chronic stress, which negatively affects health.

External stress may also lead to internal stress. Research has shown an association between external stressors and more general internal stress in racial minorities. Racial minorities often view social interactions with higher levels of anxiety due to past discrimination. It has been hypothesized that after exposure to prejudice, these people begin to actively scan their environment for other threats, and may ruminate about such experiences. There is also severe stress that comes as a result of so-called pathological bias in some members of the majority culture. In other words, racial minorities may begin to be constantly on guard and stressed.

Treatment for minority stress

In treating minority stress, one can target the societal stressors that minorities deal with, and minorities themselves. Decreasing societal mistreatment of minority groups, and thereby external stress for minorities, is an on-going, and challenging, task. Education in and acceptance of diversity, legal equality, and political support for minority groups can all help to relieve minority stress. In treating minorities themselves, it has been clinically shown that training on how to avoid thinking about instances of prejudice may teach people to shift their focus from experiences of discrimination onto more positive experiences by engaging in pleasurable or meaningful activities. And training on how to eliminate internalized stigmas through the promotion of self-acceptance is effective in reducing stress.

STRESS AND RISK OF ILLNESS
Social interaction and your health

As we have shown, it is clear that chronic stress can threaten our health. A majority of the stressors in these issues are related to psychosocial stress. This shows how pervasive psychosocial stress can be, and how important it is to treat and combat it.

Psychosocial stress and physical illness

Researchers have looked into how the number, presence, and quality of social relationships in people's lives affects their physical well-being. One study showed that socially isolated individuals are at greater risk of dying from any cause. Those with fewer social contacts also have higher chances of developing many physical illnesses, for example, cardiovascular disease. Your social status has an inverse relationship with how likely you are to develop chronic diseases. The lower an individual's social status is, the higher the chance he or she develops illnesses such as gastrointestinal, musculoskeletal, pulmonary, and renal diseases.

Psychosocial stress and mental illness

Psychosocial stress also increases our risk for developing mental illness. One study found that people who experienced conflict with coworkers or supervisors in the last five years had a higher chance of being diagnosed with a mental illness. It has also been found that people who experience stress that involves social rejection develop depression three times faster than people who experience stress that does not involve social rejection. People who are not clinically depressed have more depressive symptoms if they have friends and family who make many demands, criticize them, and create conflicts. Especially in wives, conflict between married couples results in greater psychological distress and depressive symptoms. Married couples that are unhappy have a ten to twenty-five times higher chance of developing clinical depression.

Psychosocial stress can also aggravate or deter recovery from existing psychopathological disorders. Individuals who have recovered from depression or bipolar disorder are twice as likely to relapse if they experience familial tension. Similarly, those recovering from eating disorders have a higher chance of relapsing if family members make critical comments, are hostile, or are over-involved. Studies have also revealed that social deprivation can lead to or exacerbate anxiety, depression, anger, cognitive disturbances, perceptual distortions, obsessive thoughts, paranoia, and psychosis.

Stress and coping strategies

The ways that people cope with stress affect how stress impacts their health. While there are many different coping styles, we will focus on three of the major distinctions that are often made when describing them: active vs. avoidant, problem-focused vs. emotion-focused, and responsive versus proactive. It is important to note that these categories are not mutually exclusive, so a coping strategy can be active, emotion-focused, and responsive.

- Active coping involves taking action to directly manage or solve a problem; while avoidance coping involves finding ways to distance, ignore, or distract oneself from a stressful situation.
- Problem-focused coping is when someone takes action that is goal-oriented and directed at changing or resolving a problem, or finding other solutions. In emotion-focused coping, individuals take action to help themselves, manage their emotions, and feel better, like talking to a friend.
- Responsive coping is when someone takes action after a stressful situation to deal with the problem and his or her emotions; while proactive coping is when someone takes action to prevent or alleviate the stress of potential stressful situations.

People who react to stress with problem-focused coping have been found to adapt more successfully to stressful situations, and to experience greater psychological functioning and perceived effectiveness. Individuals with major mental illnesses who use more problem-focused coping strategies tend to show less severe psychiatric symptoms than those who use emotion-focused or avoidance coping strategies. Emotion-focused and avoidance coping have been positively associated with maladaptive mental and physical outcomes. Many individuals with major mental illness tend to use less problem-focused coping, and more emotion-focused and avoidance coping. Using more emotion-focused coping and less problem-focused coping is also related to exacerbation of psychiatric symptoms.

Treating and combating psychosocial stress

Although social interactions are the main source of stress for most people, individuals also depend on social interactions to cope with stress. In order to treat and combat psychosocial stress and the health consequences that come with it, clinicians frequently encourage patients to develop relationships, reduce conflict in relationships, and develop healthy coping strategies. Patients usually benefit from getting involved in many different communities and form a diverse array of relationships. This is because studies have shown that having a diverse social network, such as having relationships with family, friends, coworkers, romantic partners, and people in a religious community, can reduce susceptibility to colds.

Social relationships can also provide support during times of stress. Using social support, especially from females, is an important factor that impacts the effect that psychosocial stress can have on an individual. A study that looked at the effectiveness of social support from females vs. males found that social support from females lowered blood pressure responses to stress more effectively in both males and females than social support from males. However, having any type of social support is better than not having any social support at all. Additionally, changing the way one copes with stress by adapting healthier coping strategies can also be beneficial. Active, problem-focused, and proactive coping are more beneficial for the health of individuals in many situations than avoidance, emotion-focused, or responsive coping styles.

FEMALE SUPPORT
Psychosocial support is important to combat the health consequences of stress. Women are reported to be particularly effective in this role. A study found that social support from females is more effective in lowering blood pressure responses to stress than from males.

Chapter Nine

FACING FEAR: RESILIENCE AND POST-TRAUMATIC STRESS

Life is full of challenges and adversities. When life events are unpredictable and traumatic, we start to feel that our lives are threatened and we are no longer able to maintain a sense of control. When severe traumatic events take place, our brains may lose the capacity to maintain homeostasis, and we may develop trauma-related symptoms that include re-experiencing the traumatic event, avoiding reminders of the trauma, and increased anxiety and emotional arousal. These symptoms can cause tremendous distress and impair our ability to carry out daily activities. Some personal characteristics have been found to be associated with being resilient to trauma and the ability to rapidly recover from stress. They include positive emotion and optimism, cognitive flexibility, emotion regulation skills, high levels of perceived coping self-efficacy, strong social support, altruism, commitment to a valued cause or purpose, the capacity to extract meaning from adverse situations, support from religion and spirituality, and attention to health and good cardiovascular fitness.

THE RESILIENT HUMAN
What is resilience?

There has not been a universally accepted way to define resilience to stress. According to the American Psychological Association, resilience is a dynamic process of adapting well in the face of adversity, trauma, tragedy, threats, or even significant sources of stress. Resilience emphasizes our capability to cope with adversity. It includes two dimensions: the significance of adversity, and positive adaptation. Adversity is considered to be the exposure to any risk factors associated with negative life conditions, such as poverty, disease, the death of a loved one, or the experience of a disastrous event.

On the other hand, positive adaptation is considered to be the manifested behaviors of social competence when facing adversity during a specific life stage, such as remaining calm after a disastrous event. Some argue that this definition of resilience is problematic because it does not adequately account for cultural and contextual differences in how people express resilience. Most research now shows that resilience is better understood as the opportunity for and capacity of individuals to navigate their way to psychological, social, cultural, and physical resources that sustain their well-being or protect them against risk factors.

Several neurobiological factors and systems are involved in maintaining resilience in times of adversity. For example, the hippocampus is involved in memory forming, organizing, and storing. The cingulate cortex is involved with emotion formation and processing; and the amygdala plays a key role in emotional reaction. In general, the hippocampus and anterior cingulate cortex inhibit stress-induced hypothalamus-pituitary-axis (HPA) activation, whereas the amygdala enhances glucocorticoid secretion. In response to stress, all these brain structures work together to modulate the HPA axis, the autonomic nervous system, and the immune response to release the optimal amount of corticosteroids, epinephrine, norepinephrine, and cytokines, to help to prepare the individual to react to the situation.

Cingulate gyrus
Thalamus
Nucleus accumbens
Anterior cingulate
Fornix
Septal nuclei
Hypothalamus
Olfactory bulb
Amygdala
Hippocampus
Mammillary body
Parahippocampal gyrus

THE LIMBIC SYSTEM

The limbic system plays a key role in the stress response, and is therefore also important in resilience to stress. Short-term stress response, accompanied by a reasonable release of cortisol via limbic activation of the HPA axis, provides resilience by generating responses that prepare the body for appropriate action and protect the body from harm.

Many psychosocial characteristics are associated with resilience. They include positive emotion and optimism, cognitive flexibility, emotion regulation skills, and high levels of self-efficacy—your belief in your own abilities to deal with situations. Other factors include strong social support, altruism, commitment to a valued cause or purpose, support from religion and spirituality, the capacity to extract meaning from adverse situations, as well as attention to health and good cardiovascular fitness, good sleep hygiene, healthy nutritional habits, and the capacity to rapidly recover from stress.

Positive emotions

Resilience and positive emotions are closely related. Positive emotions are not merely a by-product of resilience, but serve an important function in helping us to recover from stressful experiences, and generating adaptive responses in our coping process. Maintaining positive emotions while facing adversity can promote flexibility in thinking, problem solving, adaptive coping, as well as increased personal well-being. If we use coping strategies that elicit positive emotions, such as benefit-finding and positive thinking, humor, and optimism, we have a stronger resistance to stress as we feel we have greater control over difficult situations.

Additionally, positive emotions can have a physiological impact by improving the functioning of our immune system; and by increasing the levels of salivary immunoglobin A, a vital antibody that serves as our body's first line of defense against respiratory illnesses. Other health benefits include faster injury recovery rates and lower readmission rates to hospitals for the elderly, and reductions in a patient's stay in hospital. People with positive emotions are generally more resilient and recover faster after they suffer from a stressful event.

Resilience and depression

There are several areas that have been studied in the development of interventions that boost resilience for the prevention and treatment of depression, which are discussed in the sections that follow.

Genetics and the environment

Genetic and developmental factors play an important role in our response to stress and trauma. Genes can be influenced by environmental changes. Thus, by changing the biological and psychosocial environment, resilience may be promoted or suppressed. For example, it is critical to provide children with a supportive and loving environment that fosters healthy attachment, protects them from repeated experiences of uncontrollable stress, and provides them with ample opportunities to master life's challenges. Interventions that teach children and adults the skills needed to improve social competence, and construct and maintain supportive social networks, are likely to enhance resilience and decrease rates of stress-related depression.

Example	Negative emotion	Positive emotion
Emotion	Feeling anger, disgust, fear, or sadness	Feeling happiness, hope, joy, or love
Message	Something is wrong	Everything is fine
Impulse	Wanting to attack, escape, or hide	Wanting to explore, play, or socialize
Options	Few, limited possibilities for keeping safe	Many possibilities for allowing development
Consequence	Limits people's abilities and options of how they respond to stress	Broadens people's abilities and options of how they respond to stress

POSITIVE EMOTIONS AND STRESS
When coping with stress, if people can react positively rather than negatively, the positive emotions can help them to promote adaptive response and resilience.

Cognitive and/or psychological interventions

Several therapeutic approaches have been designed to help us to modify appraisals of threat and adversity, including training in concentration, emotion management, and self-efficacy. Interventions, such as concentration exercise and mindfulness training, teach us how to control where we direct our attention. These interventions have shown promise as treatments for depression. Emotion management training is associated with resilience and can moderate the relationship between the severity of life stress and depression. Such training is a central component of many cognitive-behavioral therapies, which are effective treatments for depression and post-traumatic stress disorder (PTSD).

Self-efficacy refers to our perceived capacity to successfully manage and recover from the demands of a stressful situation. One of the most important aspects of obtaining self-efficacy involves mastery experiences. In mastery experiences, we learn the skills needed to successfully manage a stressor, and then practice those skills, preferably with feedback, in increasingly challenging situations until we have mastered the challenge.

Neurobiological interventions

Stress and stress hormones produce both adaptive and maladaptive effects on the brain throughout our lifespan. One of the maladaptive effects of stress on the brain is stress-induced structural remodeling (see Chapter Two) in the hippocampus, amygdala, and prefrontal cortex, leading to excessive anxiety, fear, and cognitive impairment. Therefore, a better understanding of the neurobiology of resilience will lead to prevention and improved treatment for stress-related disorders such as depression and PTSD. For example, enhancing the function of the neurotransmitter neuropeptide Y, particularly for individuals who do not naturally release sufficient amounts of the neurotransmitter, might boost physiological resilience by keeping the sympathetic nervous system and HPA axis at an optimal level of activation. Similarly, developing therapeutic agents to prevent stress-induced overproduction of corticotropin-releasing hormone, which controls the body's response to stress, would likely reduce the frequency and intensity of trauma-related symptoms.

Other mediators of stress resilience that are targeted to reduce the risk of developing stress-related depression include the serotonergic, dopaminergic, noradrenergic, gamma-aminobutyric acid (GABA), and glutamatergic systems. For example, serotonergic and noradrenergic antidepressants protect against stress-induced learned helplessness in animals. These antidepressants stimulate the regrowth of hippocampal neurons that are damaged by stress. Antiadrenergic agents, like propranolol, may have a role in reducing traumatic memories.

THE RESILIENT NEUROTRANSMITTER

NEUROPEPTIDE Y
A molecular model of neuropeptide Y, a neurotransmitter produced in the brain and in the peripheral nervous system in response to stress. It helps modulate the response, reducing anxiety—and studies show that people who produce more of it tend to be more resilient to stress.

POST-TRAUMATIC STRESS STUDIES

Understanding post-traumatic stress disorder

Post-traumatic stress disorder (PTSD) can develop following a traumatic event that threatens our safety or makes us feel helpless. Any overwhelming life experience can trigger PTSD, especially if the event feels unpredictable and uncontrollable, such as war, assault, sexual or physical abuse, the sudden death of a loved one, or the experience of natural disasters.

The onset of PTSD varies from person to person. While the symptoms most commonly develop in the hours or days following the traumatic event, it can sometimes take weeks, months, or even years before they appear. PTSD can affect those who personally experience the catastrophe, those who witness it, and those who pick up the pieces afterward, including emergency workers, medical staff, and law enforcement officers. The disorder can even occur in the friends or family members of those who went through the actual trauma.

Following a traumatic event, most people experience at least some of the symptoms of PTSD, such as feeling numb, fearful, disconnected, or having nightmares. For the majority of people, those symptoms only last for several days or weeks and gradually disappear. A formal diagnosis of PTSD refers to people with symptoms lasting for more than a month and causing significant distress in daily life.

Children and PTSD

Children, especially those who are very young, can express PTSD differently than adults. PTSD symptoms in children include fear of being separated from a parent, loss of previously-acquired skills (for example, toilet training), and compulsive play in which themes or aspects of the trauma are repeated. They may

SYMPTOMS OF PTSD

There are three ways that PTSD symptoms are typically experienced:

1. Re-experiencing the traumatic event, for example flashbacks, nightmares, intense physical or emotional reactions when reminded of the event.

2. Avoiding reminders of the trauma, for example the inability to remember important aspects of the trauma, loss of interest of daily activities, feelings of detachment from others or surroundings.

3. Having increased anxiety and emotional arousal, for example difficulty falling or staying asleep, problems concentrating, outbursts of anger, vulnerability to being startled.

suffer new phobias and anxieties that seem unrelated to the trauma. They may also experience aches and pains that have no apparent cause.

PTSD causes and risk factors

Traumatic events are more likely to cause PTSD when they involve a severe threat to life or personal safety; the more extreme and prolonged the threat, the greater the risk of developing PTSD. Intentional, human-inflicted harm—such as rape, assault, and torture—also tends to be more traumatic than impersonal accidents and disasters. The extent to which the traumatic event is unexpected, uncontrollable, and inescapable, also plays a role in the development of PTSD. Other risk factors include personal and family history of traumatic events, or mental illness, and a lack of support or coping skills after the trauma.

Genetic and developmental factors play an important role among the risk factors of PTSD. DNA studies have found that genetic factors shape the regulation of the stress response and reactivity of the sympathetic nervous system, HPA axis, neuropeptide Y system, and the serotonergic system. The best-studied gene-environment interaction involves a naturally occurring variation in the promoter of the human serotonin transporter gene, which may be specifically associated with an increased risk of depression following exposure to childhood maltreatment. Repeated episodes of overwhelming stress during infancy and childhood can lead to "learned helplessness" and can cause exaggerated emotional behavior when faced with stress, even into adulthood.

Treatment for PTSD

Recovery from post-traumatic stress disorder (PTSD) is a gradual, ongoing process. Treatment for PTSD relieves symptoms by helping people deal with the trauma they experienced rather than avoid the trauma and any reminder of it. Treatment encourages individuals to recall and process the emotions and sensations they felt during the original event. In addition to offering an outlet for emotions they have been bottling up, treatment also helps individuals to restore their sense of control and reduce the powerful hold the memory of the trauma has on their lives. Treatment for PTSD includes trauma-focused, cognitive-behavioral therapy, family therapy, medication, and eye movement desensitization and reprocessing (EMDR). There are also self-help interventions to help people recover from PTSD, such as reaching out to others for support, avoiding drugs or alcohol, spending time in nature, and overcoming the sense of helplessness (for example, helping others, or taking positive actions).

For people who have family members suffering from PTSD, they need to first take good care of themselves and get extra support. It is common for family members of PTSD patients to ignore their own needs, which leads to them burning out. It is also helpful for them to learn as much as they can about PTSD to facilitate the patient's recovery.

Resilience and vulnerability in the war zone

Studies of World War II combat veterans show that soldiers with certain characteristics and experiences are more vulnerable to developing post-deployment stress injuries and substance-use disorders.

Pre-deployment

A number of pre-deployment factors can contribute to increased vulnerability to stress. They include uncertainty, routine changes, a struggle to make arrangements, and concern about themselves and family members. Single parents, individuals in the reserve forces, and those who have no previous deployment experience extra stress.

In general, the level and quality of pre-deployment training is an important predictor of post-deployment stress. To reduce combat-related stress reactions, the US Army has developed an extensive resilience training program called "Battlemind" to develop and boost a soldier's will and spirit to fight and, ultimately, to win in combat.

During deployment

According to reports, the most important factors that lead to resilience during deployment are cohesion, bonding, and buddy-based support within the military unit.

There are also several risk factors for increased vulnerability to stress during deployment, including the severity of exposure to combat, and the degree of life threat or perceived life threat. In the US, The Mental Health Advisory Team (MHAT) IV was established by the Office of Army Surgeon General to assess soldier and marine mental health and well-being, and examine the delivery of behavioral health care during the Iraq War. A survey conducted by the MHAT IV indicated that "[longer] Deployment length was related to higher rates of mental health problems and marital problems." Overall, risk factors identified in the survey include combat exposure, deployment concerns, multiple deployments, deployment length, pre-existing behavioral health issues, anger, and marital concerns.

The experience of killing

Feelings of responsibility and guilt may worsen some veterans' post-combat stress effects. Participation in war-zone violence can result in post-military violence to self, spouse, or others. Those veterans who reported that they had killed in combat tend to have higher PTSD scores. Scores were even higher for those who said they were directly involved in war atrocities. Additionally, one study of suicide attempts among Vietnam combat veterans showed that guilt about actions in combat were the most significant predictor of suicide attempts.

Coping styles in the war zone

Characteristics, including high sociability, thoughtful and active coping styles, and a strong perception of one's ability to control one's destiny, are found to be associated with greater resistance to traumatic stress. In survivors of war and disaster, people who cope by

EFFECTS OF WAR

War veterans who reported they had killed other people and were involved in atrocities tend to have high PTSD scores, and possibly suffer from a greater risk of suicide. Goya's famous painting of the *Execution of the Defenders of Madrid* (1808) illustrates the horrors of war.

cooperating with others tend to escape trauma and post-trauma effects. More severe post-trauma responses tend to appear among people who freeze and dissociate, and in those who react in a "Rambo" fashion by responding with isolated, impulsive action. In a study of 750 Vietnam veterans who had developed neither PTSD nor depression after being held as war prisoners, ten elements were considered critical characteristics of resilience: optimism, altruism, a moral compass, spirituality, humor, a role model, social supports, head-on confrontation of fear, a mission, and training.

COMPONENTS OF RESILIENCE
How can I stand up to stress?

Resilience, like most psychological theory, is an abstract concept. It is frequently helpful to be able to describe and measure the elements included in the concept. The Connor-Davidson Resilience scale (CD-RISC) is commonly used to estimate resilience in a population. This scale comprises twenty-five items and has a total score that can range from 0–100, with higher scores reflecting greater resilience. The CD-RISC is a useful measure of resiliency, and can indicate accurately those with more and less resilience. The twenty-five items of the CD-RISC can be grouped into five resiliency subscales, which include perceived competence, trust in one's instincts, tolerance of negative effects, positive acceptance of change, and spiritual influences.

Perceived competence is similar to optimism, and indicates the expectancy that we can effectively interact with the environment and achieve good outcomes in life. Perceived competence is related to enhanced well-being, better stress management, and more effective self-regulation. It focuses on an individual's behavior in achieving success, whereas with optimism, the expected positive outcome does not necessarily depend on the action of the person. Yet, perceived competence alone is not sufficient for success. In many situations, success is more affected by personal competence than optimism.

Trusting in one's instincts is usually seen as being automatic, fast, and practically "thoughtless" since it does not require analysis or deep thinking. This intuitive behavior has great evolutionary benefits by permitting humans to react quickly and effortlessly to things that threaten or endanger them. However, trusting your instincts may not be always be beneficial in all situations. For example, your brain is likely to bring up the most recent, memorable, and vivid idea or experience one has been exposed to when deciding how or whether to react to a situation.

EXPERIENCING THE STRESS OF WAR
Some individuals are more resilient to trauma than others. Soldiers with a positive outlook, moral compass, and cognitive flexibility are more resilient to adverse experience in war.

Quick and instinctual reactions accompanied by mental processing are frequently advantageous.

Age and experience can influence what people think contributes to resilience. Older adults tend to credit their resilience to their "acceptance and tolerance of negative outcomes"; while younger adults tend to credit their resilience to their ability to engage in "problem-focused active coping."

Spiritual beliefs may also influence a person's outlook on the world. Spirituality offers solace in turbulent times, and provides support from a like-minded community, which can be beneficial to health.

ELEMENTS THAT PROMOTE RESILIENCY

Resiliency to stress is associated with at least six psychosocial factors: active coping styles, regular physical exercise, a positive outlook, a moral compass, social support, and cognitive flexibility.

Active coping styles focus on problem-solving and managing emotions that accompany fear and stress; active coping is related to emotional well-being. In contrast, people who use passive coping styles, including denial and avoidance of problems, substance abuse, and resignation tend to be depressed. Patients with PTSD are frequently helped if they are able to actively face their traumatic triggers and fears in a stepwise approach.

Regular physical exercise is actually a type of active coping that diminishes negative emotions caused by stress. Regular exercise has been shown to improve clinical depression in adults, build physical and emotional hardiness, lift mood, and improve memory. Through exercise, the human body releases endorphins and serotonin precursors, attenuates basal HPA axis activity, and promotes the expression of neurotrophic and neuro-protective factors. These exercise-induced neurotrophic factors include nerve growth factor and brain-derived neurotrophic factor. This factor is important because it stimulates neurogenesis in the hippocampus and appears to improve learning and memorization. Thus, exercise appears to increase brain plasticity and enhance ability to learn from and adapt to stressful situations.

A positive outlook Having a positive outlook means using a flexible thinking style to enhance optimism, decrease pessimism, and embrace humor. Depressed individuals tend to view their problems as permanent and pervasive. Humor and positive emotions make it possible to see the lighter side of a difficult situation. Possessing a moral compass is defined as having developed meaningful principles and having put them into action through altruism. Altruism benefits both the person who practices it and the person who receives it. People who help others perceive themselves as necessary and derive fulfillment from that perception. This phenomenon, known as "required helpfulness," was first described during World War II when those who cared for others after bombardments suffered less post-traumatic psychopathology than those who did not. Some individuals find healing in a "survivor mission" after personal tragedy by helping others cope with the same problem they faced.

Social support can reduce risk-taking behavior, encourage active coping, decrease loneliness, increase feelings of self-worth, and help a person put problems into perspective. Individuals with strong social support tend to be more resilient than those without. A lack of social support correlates with depression, stress, and increased morbidity and mortality during medical illness. On the other hand, people who feel connected to someone they respect can learn to manage stress by mimicking the behavior of that person, and benefit from the experience of their mentor. Many resilient adults credit a parent, grandparent, or other role model for teaching them to act honestly and inspiring them to be strong.

Cognitive flexibility, or "cognitive reappraisal," is the ability to positively reframe negative events. Early life exposure to severe trauma can cause long-term damage, while exposure to milder stressful events may build resilience. Individuals who successfully overcome adverse events usually find some meaning in their tragedy by positively reframing the event. Psychiatrist and Holocaust survivor Viktor Frankl wrote about the importance of "meaning making." Despite suffering in Nazi concentration camps, Frankl wrote that he gained the opportunity to exercise inner strength and be "brave, dignified and unselfish." Neuroimaging studies indicate that individuals who use cognitive reappraisal to deal with adversity have a strong control over their emotions. They can modify their reaction to stress by activating the prefrontal cortex, which then modulates the amygdala's response to the situation.

EXERCISE AND STRESS
Regular exercise may increase brain plasticity and enhance the ability to learn from and adapt to stressful situations.

STRESS INOCULATION AND POST-TRAUMATIC GROWTH
Building resilience and coping strategies

Stress inoculation training is a cognitive behavioral treatment for post-traumatic stress disorder (PTSD), aiming to help people gain confidence in their ability to cope with anxiety and fear stemming from reminders of their trauma. Similar to how vaccinations help individuals build a resistance to infections, stress inoculation training allows individuals to develop resiliency to trauma and stress. Immunity to stress is not specific to the type of stressor first encountered; early exposure to manageable stress appears to enhance resilience to many adverse experiences. During the training, a therapist helps clients become more aware of what things trigger fear and anxiety. Additionally, clients learn a variety of coping strategies that are useful in managing anxiety, such as muscle relaxation and deep breathing. Other factors that contribute to

resilience to stress or trauma may include social support, self-assurance, and the ability to react appropriately in the face of fear.

Post-traumatic growth
In the aftermath of trauma, many resilient people experience what psychologists Richard Tedeschi and Lawrence Calhoun have termed "post-traumatic growth" (PTG)—positive personal growth that can

---

STRESS INOCULATION TRAINING
The armed forces undergo training regularly. Some experience stress inoculation training, where therapists help people to gradually increase their coping skills in high-stress situations. In the final stage, participants face mock, real-life adversity. The training helps participants to deal with the emotional impact of war, as well as show resilience during testing times. This helps combatants to fulfill their roles more effectively.

result from a traumatic event. People who are open to changes of life, and approach difficulty actively tend to experience PTG. They usually undergo significant life-changing psychological shifts in feeling and thinking, which contribute to a meaningful personal process of change. The two psychologists developed a PTG inventory that measures positive outcomes across five categories:

FIVE POSITIVE OUTCOME CATEGORIES
1. New possibilities
2. Personal strength
3. Relating to others
4. Appreciation of life
5. Spiritual change

For example, people who experience PTG may become more open to new possibilities and opportunities that had not presented themselves before the traumatic event. Many people with PTG may also develop an altruistic sense of mission or become more compassionate to others because PTG forces them to focus on questions about wisdom, virtue, and values, thus bringing true meaning to their lives.

However, Tedeschi and Calhoun also point out that, regardless of the potential for positive psychological changes, PTG neither diminishes the distressing nature of the event, nor lessens feelings of pain and loss. To move toward PTG, people have to go through a phase of intense reflection. A person has to cope with the feelings of loss, anger, and other emotional pain caused by the trauma before he or she can begin reflecting on and accepting opportunities for change and growth.

Anti-fragility
Sharing concepts with stress inoculation and post-traumatic growth, anti-fragility describes the concept of not only withstanding and being resilient to shock and trauma, but of also improving and growing stronger due to stressors. The anti-fragility concept differs from growth through resiliency in that it is due to the traumatic event that the individual becomes more resilient than normal and, thus, is more able to cope with the negative effects of future stressors.

Animal studies have demonstrated this concept by showing that early experiences with successful behavioral control over stressful events induce neuroplasticity in the prefrontal cortex, which appears to protect the animal from some of the negative effects of future uncontrollable stress. The environment plays an important role in how anti-fragile an individual is. Anti-fragility is promoted by an enriched environment and consistent supportive maternal care that provide an atmosphere that fosters exposure to new experiences and the mastering of challenges. Negative and positive neurobiological and behavioral consequences of parental care can even be transmitted across generations, possibly due to changes in the environment within cells that affect gene replication activities. As approaches to enhancing resilience are still in the experimental stage, advances in our understanding of resilience will come from more research on complex interactions between genetic, developmental, neurobiological, psychosocial risk, and protective factors.

HOW TO DEVELOP RESILIENCE

Active coping styles	Focus on problem-solving and managing emotions that accompany fear and stress.
Regular physical exercise	Take regular exercise; this is a form of active coping that diminishes negative emotions caused by stress.
A positive outlook	Build a flexible thinking style to enhance optimism, decrease pessimism, and embrace humor.
Possess a moral compass	Develop meaningful principles and put them into action through altruism.
Social support	People who feel connected to someone they respect can learn to manage stress by mimicking the behavior of that person, and can benefit from the experience of their mentor.
Cognitive reappraisal	Find meanings from negative events by developing the ability to positively reframe them.

Chapter Ten

THE MIND–BODY MEDICAL EQUATION AND PUBLIC HEALTH

In this book we have endeavored to review the state of science with regard to stress and its effects on illness, in particular, the twenty-first century crisis of stress-related chronic diseases that the world now faces. The phrase used to describe these diseases is "non-communicable" (NCD) setting them apart from the "communicable" infectious diseases like HIV, tuberculosis, malaria, and Ebola. We started with an exploration of how the brain, heart, and immune systems manage stress. We then focused on stress effects on sleep, women's health, nutrition, and social experience, before examining the condition that is most closely tied to stress as an etiology—post-traumatic stress disorder. We are also interested in the practical significance of this new knowledge for the new world order of health-care reform, which is being designed to do a better job of managing chronic illnesses. This is where the significance of enhancing resilience for linking up clinical medicine and public health must take center stage throughout the world if we are to meet the sustainable development goals laid out by the United Nations in 2015. This is why our common understanding of the science of stress is so important.

THE CHALLENGE TO HEALTH
How non-communicable diseases affect our health

There are many intersections between medicine and society. What is dysfunctional in society will find its way into the mission of the medical care system and an over-extended medical system will tax our societal resources. It works both ways. This creates a situational vulnerability for all of us and for society. Stress-related challenges that create dysfunction will directly produce individual, family, and community suffering as well as societal costs.

In our time, there is a particularly difficult stress-related challenge that needs to be more effectively addressed in public health as well as clinical practice. We can no longer afford to keep the medical school separated from the public health school. Stress-related chronic non-communicable diseases (NCDs) (heart diseases, chronic lung diseases, diabetes, arthritis, and neuropsychiatric diseases) continue to plague primary care practitioners who often can only slow progression of the disease in question, leading to enormous suffering and contributing to the ballooning of health care costs to almost 18 percent of the gross domestic product (GDP) in the United States. NCDs represent the most important global health challenge of this century in terms of disease burden and mortality.

In earlier chapters we learned that chronic or massive stress experiences will predispose an individual to the metabolic syndrome characterized by high blood pressure, high blood cholesterol, reduced ability to process glucose, and obesity.

Adversity in childhood
One pathway to such a vulnerable position emerges when children experience adversity in childhood. These children suffer dearly from what is now known as toxic stress in the form of abuse or neglect, which will often lead to a chronic illness that will be costly in terms of personal distress as well as societal cost.

Developmental science now suggests that early childhood toxic stress creates a shared vulnerability for the intractable problems we face in health, education, and criminal justice. Thus, our stress-related chronic illness challenge shares a common root with the specter of failing schools and the horror of overcrowded prisons. Of course the suffering and societal cost of ignoring this situation are enormous.

Globally, we also face the challenges of returning combatants from countries at war, who suffer from post-traumatic stress, especially when complicated by traumatic brain injury. This is a crisis not only for the wounded veterans themselves, but also for their families and future generations in their families, especially when complicated by traumatic brain injury. And then there is the tsunami we will face

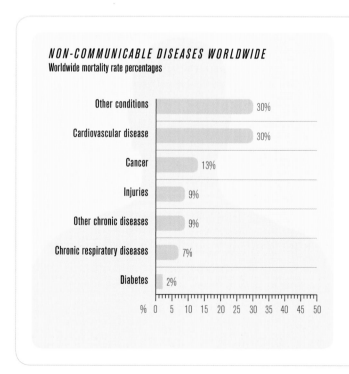

NON-COMMUNICABLE DISEASES WORLDWIDE
Worldwide mortality rate percentages

Other conditions	30%
Cardiovascular disease	30%
Cancer	13%
Injuries	9%
Other chronic diseases	9%
Chronic respiratory diseases	7%
Diabetes	2%

% 0   5   10 15 20 25 30 35 40 45 50

with regard to the stress of an ever-increasing number of family caregivers responsible for the care of the growing elderly population.

Climate change

Nowadays, the world is waking up to the potential ill effects of climate change. Consider the direct and indirect stress effects of natural disasters such as floods, hurricanes, and fires, along with the human and societal toll of displacement and migration. Food shortages and the lack of potable water from drought will also cause extreme stress. And the direct health effects of heat stroke must also be taken into account. All of these factors will put the victims of climate change at grave risk for the illness-promoting effects of severe allostatic loading.

All of these stress-laden conditions set the stage for the non-communicable disease crisis we are facing in this century. This is because NCDs are the downstream clinical repercussions of the stress-related metabolic syndrome described above. And the personal and societal costs secondary to NCDs are enormous. The figures below depict the

seriousness of NCDs based on mortality data and the so-called Disability Adjusted Life Year (DALY) measure, which signifies not only the death rate due to specific illnesses, but also the amount of time lost to illness-related disability.

The NCD challenge to the budget

In recognition of the growing importance of NCDs, the World Economic Forum commissioned the Harvard School of Public Health (US) to study and produce a report focused on cost of these illnesses. The researchers reached these conclusions:

▸ Heart diseases, chronic lung diseases, cancer, diabetes, and mental illnesses represent a cumulative output loss of $47 trillion dollars, roughly 75% of the global GDP in 2010.

▸ Costs are very high and projected to grow significantly by 2030, with huge productivity losses due to death and disability.

NON-COMMUNICABLE DISEASES

NCDs constituted more than 36 million deaths (60%) worldwide in 2005 according to the WHO; 80 percent of deaths occurred in low- and middle-income countries (LMICs).

Source: Courtesy of WHO

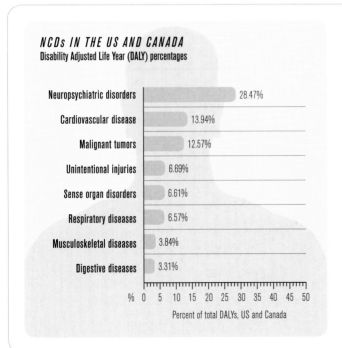

NCDs IN THE US AND CANADA
Disability Adjusted Life Year (DALY) percentages

Neuropsychiatric disorders	28.47%
Cardiovascular disease	13.94%
Malignant tumors	12.57%
Unintentional injuries	6.69%
Sense organ disorders	6.61%
Respiratory diseases	6.57%
Musculoskeletal diseases	3.84%
Digestive diseases	3.31%

% 0 5 10 15 20 25 30 35 40 45 50

Percent of total DALYs, US and Canada

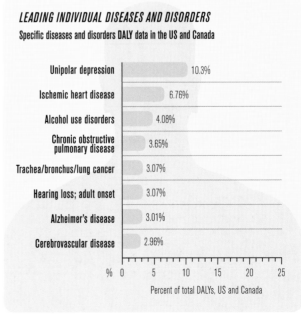

LEADING INDIVIDUAL DISEASES AND DISORDERS
Specific diseases and disorders DALY data in the US and Canada

Unipolar depression	10.3%
Ischemic heart disease	6.76%
Alcohol use disorders	4.08%
Chronic obstructive pulmonary disease	3.65%
Trachea/bronchus/lung cancer	3.07%
Hearing loss; adult onset	3.07%
Alzheimer's disease	3.01%
Cerebrovascular disease	2.96%

% 0 5 10 15 20 25

Percent of total DALYs, US and Canada

STRESS AND OUR HEALTH
How do we promote health and prevent illness?

The healthy human psychological environment has been determined for us through eons of evolutionary change. It consists primarily of the basic experience of secure attachment—a non-stressful relaxed state. The health of all mammals, especially primates, is built on this foundation as the father of attachment theory, the English psychiatrist, John Bowlby (1907–90), once proposed. Stress is pervasive as it is a condition of living with ubiquitous separation challenges to our attachments. The term *stress* refers to the overall concept of your physiological response to environmental triggers called *stressors* that kindle the threat of being separated. This is why a key ingredient in promoting your health involves your social attachments and support. With this in mind, the prevention and treatment of the harmful effects of stress on your health and well-being is vital and should be an important feature of your personal integrative health-care plan as well as the integrative components of overall health-care reform, particularly in light of the NCD crisis reviewed earlier.

Childhood maltreatment is the antithesis of long-term health promotion. Such early life stress will result in a significant and graded increase in the risk for illness as reflected in clinically relevant elevations in inflammatory markers such C-reactive protein, a known risk factor for ischemic heart disease. Indeed, many researchers now see stress-related chronic immune system activation as the underpinning for chronic diseases. Childhood toxic stress will also make more likely an increase in fibrinogen levels, placing blood vessels at risk and in white blood cell count 20 years later in adulthood, even after adjustment for other risk factors and potential mediating variables. Indeed adverse childhood events (ACEs) including abuse, socioeconomic disadvantage, and social isolation, are associated at age 32 years with the presence

of major depression and the four major components of metabolic syndrome (obesity, hypertension, high cholesterol, and insulin receptor hyposensitivity). The metabolic syndrome as stated earlier is endemic in our society and is major precursor to our major stress-related chronic diseases.

The majority of our visits to health-care providers in industrialized countries are related to stress and its manifestations. Stress has a profound adverse influence on our physical and mental health, on our performance and efficiency in the workplace, and on education of our young people. Stress of a psychosocial nature is linked to biological system stress at the cellular level leading to what is called oxidative stress. Oxidative stress creates metabolic by-products that are harmful to our cells making metabolism inefficient, and it is this linkage that leads to the uncovering of our physiological disease vulnerabilities. These vulnerabilities are engendered by both our genetic endowments and our environmental happenstance.

Experience is perceived as stressful when it signifies to you a discrepancy between what "*should*" be (the set goal variable of secure attachment) and what actually is (the actual variable). This implies that there is a disturbance in the normal state of physical and mental affairs that will need your attention and action. In essence this is an alarm system designed to produce in you neurophysiological activation and arousal, which is necessary when danger is perceived. In this context stress can be defined as a state of disharmony or threatened stability. Biochemical (neurotransmitters, peptides, steroids), physiological (heart rate, blood pressure), and behavioral (anxiety, depression, tension) concomitants of stress may co-mediate a disease response, which precipitates down to the level of cellular oxidative stress as described in previous chapters.

Musculoskeletal system
Chronic stress can cause severe headaches, neck aches, and other muscular problems.

Cardiovascular system
Chronic stress can worsen heart function resulting in heart attacks and arrhythmia becoming more common.

Respiratory system
Chronic stress can cause exacerbation of bronchitis and emphysema, and can lead to smoking, which worsens lung diseases.

Nervous system
Chronic stress strengthens our fear response and weakens the brain regions that control our stress response systems. This can cause depression, stroke, and neurodegenerative disorders.

Endocrine system
Chronic stress can lead to adrenal overproduction of cortisol, thyroid gland changes, and liver outpouring of glucose, all making the metabolic syndrome including diabetes more common.

Gastrointestinal system
Chronic stress affects the brain–gut axis leading to digestive diseases such as gastroesophageal reflux, inflammatory bowel disease, and irritable bowel syndrome.

Cancer
Chronic stress may cause changes in tumors that can increase the risk of cancer.

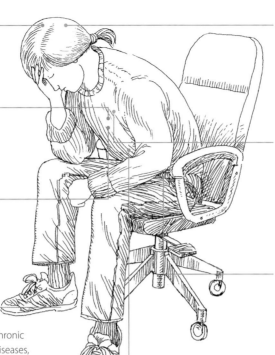

STRESS-RELATED DISEASES
The long-term effects of stress can lead to chronic NCDs such as heart diseases, chronic lung diseases, diabetes, arthritis, digestive diseases and neuropsychiatric diseases. Allostatic load contributes to the development of the metabolic syndrome and can lead to the onset or exacerbation of these disabling diseases with high mortality and morbidity.

As hinted at earlier, attachment theory can inform our understanding of stress and provide a hint as to how we can promote health. Attachment can be understood as a system that behaviorally regulates interpersonal synchrony (attunement) while stress can be defined as an asynchrony in the interaction (misattunement). When interpersonal social harmony is re-established (re-attunement) there is recovery from stress and successful coping has occurred.

In one 35-year study researchers found that 100 percent of those who reported lack of warmth and poor closeness to both parents were diagnosed with diseases such as ischemic heart disease, hypertension, ulcer disease, and alcoholism in middle age. Only 45 percent of those who described warmth and closeness had midlife diagnoses like these. Your resilience and propensity to health may be strongly related to your parental nurturance and secure attachment in childhood; but nevertheless there are reasons for all of us to hope that as adults we can enhance our feelings of secure attachment and reap the benefits for our health.

THE INADEQUACY OF MEDICAL TREATMENT

Doctors can do more to help their stressed patients and thereby alter the path toward the metabolic syndrome and NCDs. Pressed for time during short routine visits, doctors may find it difficult to advance the causes of mind–body and integrative health promotion and illness prevention. Yet this is the foundation for a sound prevention plan. Prevention focuses on keeping healthy and at-risk people healthy. In addition to diagnosing and managing acute exacerbations of stress-related chronic diseases with medications and procedures, doctors can apply good public health principles in the clinic setting. They will benefit patients, families, and their communities by providing prevention advice for mind–body self care to help ill patients avoid future acute problems. But they can also take advantage of opportunities to avoid future NCDs by appraising the stress and resilience of check-up patients who are presently healthy and by advocating mind–body self-care practices.

THE MIND–BODY INTEGRATIVE MEDICINE EQUATION
A guide to improve health and prevent illness

Based on the material presented in this book a pragmatic Mind–Body Integrative Medicine Equation may be proposed. This illness index equation can serve as a guidepost to all of us for health promotion and illness prevention strategies. Here, the underappreciated work of the American psychologist George Albee (1921–2006) and others bears mention.

These theorists synthesized the concepts needed to promote health and prevent illness not only in mental health but also with regard to stress-related illness of any kind. The best opportunity for us to make strides in all aspects of health promotion and illness prevention is by reducing the numerator, namely stress effects on physical and mental vulnerabilities, and by enhancing the denominator, namely healthy behaviors, social competencies, supportive resources, and social connectedness.

By doing this you can, in an estimated way, take stock of your health and illness prospects by considering your personal numerators and denominators. You would do well to bring your estimates to your primary care visits and engage with your physicians in a discussion of how they might improve your chances of promoting your health and performance and preventing illness. Disease will emerge in vulnerable individuals when psychosocial stress produces the level of metabolic wear and tear that will unleash cellular oxidative stress.

Maintaining stability through change
How you respond to the wear and tear that your body experiences will vary, based on factors that encourage resilience and factors that favor vulnerability. As we plan our strategies for health promotion and illness prevention, we need to keep in mind how we might enhance our resilience and reduce our stress vulnerability. Only in this way will the crisis of chronic stress-related diseases begin to abate.

Can integrative mind–body approaches reduce allostatic loading, reversing these negative aspects of the stress response? The answer appears to be "*yes*" if we can reduce the numerator—stress and its negative

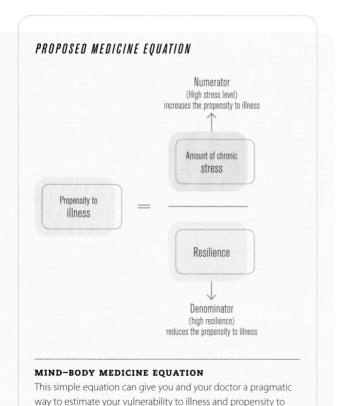

PROPOSED MEDICINE EQUATION

Numerator
(High stress level)
increases the propensity to illness

Amount of chronic **stress**

Propensity to **illness** =

Resilience

Denominator
(high resilience)
reduces the propensity to illness

MIND–BODY MEDICINE EQUATION
This simple equation can give you and your doctor a pragmatic way to estimate your vulnerability to illness and propensity to health. By learning ways to reduce chronic stress (numerator) and build resilience (denominator) we stand a better chance of lowering the illness quotient and remaining healthy.

THE BENEFITS OF EXERCISE

Regular exercise can work wonders for your own personal mind–body medicine equation. Exercise changes your brain and appears to strengthen those areas that serve to dampen the amygdala fear center's excitatory overworking of the brain as a whole. For example, aerobic exercise has been shown to increase the neurohormone brain-derived neurotrophic factor resulting in new neuron birth in the hippocampus. Running appears to increase local inhibitory mechanisms in the hippocampus to neutralize the amygdala's overstimulation and to reduce stress-related oxidative stress, thereby improving cellular mitochondrial reserve. The effect of such changes are to reduce stress and build resilience leading to a better chance of remaining healthy.

effects on health—and increase the denominator—resilience and its positive effects on health. Mind–body approaches like meditative techniques that elicit the relaxation response and mindfulness, cognitive skills training, social support, and optimism, all have the potential to build resilience through mechanisms of reward and motivation, decreased responsiveness to fear, and adaptive social behavior.

Lowering the distress numerator

The relaxation response (RR) as originally described by American doctor Herbert Benson (1935–) in the 1970s is a self-induced stimulus that breaks the hold of a stress response, thus lowering the numerator in our equation. It is defined by a set of integrated physiological adjustments that occurs whenever you enter into a repetitive mental or physical activity and passively ignore distracting thoughts. This response can be encouraged through breathing techniques, mindfulness meditation, certain forms of prayer, tai chi, qigong, yoga, and autogenic training, and is associated with improvement in your breathing efficiency (decreased oxygen consumption and

carbon dioxide elimination), and reductions in your metabolism through lowered heart rate, arterial blood pressure, and respiratory rate. The relaxation response can be thought of as the physiological counterpoint to the stress or fight-or-flight response. The relaxation response has been shown to reduce stress-related anxiety and depression symptoms, and may benefit several conditions, including high blood pressure and cardiovascular disease.

Meditation creates a state of relaxation commensurate with physical markers reminiscent of the physiological profile that accompanies the experience of being securely attached to a loved one. This is why the relaxed meditative state is often said to establish an oceanic feeling of connection in the present, reminiscent of the attunement between mother and child. During meditation, the predominance of the parasympathetic nervous system is reflected in lower heart rate and blood pressure, decreased skin conductance, increased belly respiratory amplitude, decreased chest respiratory rate, and increased frequency of heart rate variability.

Mindfulness

Mindfulness is that aspect of consciousness that allows selective attention to one mental focus resulting in an increased clarity of awareness. Mindfulness-based stress reduction (MBSR) is an eight-week course, which promotes relaxation while also enhancing selective attention, and changing cognitive appraisal to a healthier coping stance. It does this by reducing the individual's focus on self-related and distorted cognitions and goals, and by strengthening focus on the present moment, which is re-perceived as less threatening. As a result, those practicing mindfulness may benefit from less autonomic reactivity in the setting of social separation challenge because they perceive less threat and less hostile intent. Perceived stress is reduced with mindfulness practice and perceived sense of control may be increased, ironically by reducing one's perceived need to be in control.

Increasing the resilience denominator

In order to enhance your resilience, the heeding of standard health advice is still essential. Good sleep hygiene involves attempting to get seven to eight hours of sleep a night. Good nutrition still includes elements of the standard food pyramid, with attention paid to portion control and choosing foods with a low glycemic index as elucidated in Chapter Seven. And a good exercise regimen requires integration in your weekly lifestyle in order to develop at least thirty minutes a day of some exertion. All three of these lifestyle choices along with the avoidance of overuse of alcohol, and of any use of tobacco or other recreational drugs, will serve you well in increasing the denominator in your equation. It should be noted that proper exercise, diet, and sleep all could bolster the capacity of your cell's mitochondria to process energy efficiently.

Cognitive behavioral therapy

Learning cognitive skills allows you to substitute positive thoughts for the negative ones that provoke anxiety and depression. It usually involves six to twelve weekly individual sessions with a therapist; there are also many helpful online versions of cognitive behavioral therapy (CBT). In CBT,

RESILIENCE CAPACITY

Your resilience consists of five major capacities:

1. Your capacity to experience reward and motivation nested in a personal disposition marked by optimism and high positivity about life;

2. Your capacity to control responses to fear so that you can continue to be effective through active coping strategies despite feeling fearful;

3. Your capacity to use adaptive social behaviors to secure support through bonding and teamwork and to provide support through altruism;

4. Your capacity to flexibly use cognitive skills to reinterpret the meaning of negative events in one's life in a more positive light;

5. Your capacity to integrate a sense of purpose in life along with a moral compass leading to a life filled with meaning and spiritual connectedness.

participants record automatic thoughts and the automatic feelings that tend to accompany them in homework diaries. They especially note situations that cause anxiety and depression. They then propose alternative behaviors that relieve these mood states, which the CBT students then practice. CBT has been shown to be effective for anxiety and depression and also to have some success in patients with chronic, stress-related medical illnesses.

Mindfulness-Based Cognitive Therapy (MBCT) combines MBSR with CBT and targets depressive ruminations and negative thinking. It uses the process of "de-centering" in which thoughts are experienced as evanescent mind events rather than accurate depictions of reality. The use of MBCT has been shown to be as effective as antidepressant medication for prevention of depression relapse.

Social relationships influence your health promoting behaviors such as adherence, diet, and exercise. These factors in turn influence health-relevant biological states resulting in health propensities or disease risks.

Social relationships also directly impact the health relevant biology of your brain and contribute to your physical and mental status. Social resources will influence your appraisal of perceived stress and, if attachment security is thereby strengthened, your disease vulnerability will be lowered.

Social stressors

Social stressors are separation challenges and include: isolation, marital and family discord, social conflict, job strain, unemployment, physical morbidity, retirement, and social inequalities. The effects of social support in reducing your risk of disease and improving your health have been shown in studies of cardiovascular disease, irritable bowel syndrome (IBS), and anxiety and depression among other diseases.

RESILIENCE FACTORS

▸ Social support/pro-sociality
▸ Cognitive skills
▸ Positive psychology
▸ Spiritual connectedness
▸ Exercise
▸ Nutrition
▸ Sleep hygiene
▸ Healthy habits

The placebo effect

When we truly believe in the ability of the doctor and the treatment to heal us, better health may ensue. This expectation of a future return to wellness has the power to stimulate the brain reward/motivation circuitries and this can reduce stress. In this way, belief and positive expectation can diminish disease and provide solace, leading to what has sometimes pejoratively been called the *placebo* response. This re-found state of security and attunement could result in a general reduction not only in stress-related neural, vascular, and immune overactivation syndromes but also in specific up-regulation of brain reward mechanisms. This ultimately would reduce oxidative stress and provide a fertile field for a return to health.

The flip side of the positive placebo response is the negative *nocebo* response. Here, a negative expectation of the future can precipitate negative emotions that can lead to a cascade of stress-related illness. As we reviewed in Chapter Three, the *nocebo* can be especially devastating in the context of heart disease, where negative emotional states such as depression and anxiety are associated with adverse cardiac events. By contrast, positive psychological states such as conscious positive expectation or optimism are associated with better cardiac outcomes in subjects with and without histories of cardiac disease. This optimism has now been shown to predict improved global health status, reduced all-cause mortality, and improved mood.

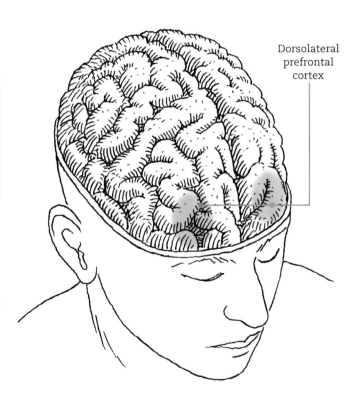

Dorsolateral prefrontal cortex

PLACEBO EFFECT IN THE BRAIN
Experiments have shown that the area most heavily involved in instigating the placebo effect is the dorsolateral part of the prefrontal cortex. It is involved in complex functions such as expectation and belief, and can directly cause the release of endogenous opioids (natural painkillers).

IMPROVE OUR HEALTH, WELLNESS, AND PERFORMANCE

How to cope with stressors

Simply by inverting the numerator and denominator in the mind–body medicine equation, you can reach an integrative health index version of it. Your propensity to health and optimal performance is defined by a relationship between the stress of separation challenges and genetic vulnerability (as denominator) and attachment solutions, which help us to cope better with stressors (resilience) and genetic endowment as numerator. Mind–body self-management—the ability to elicit the relaxation response, cognitive skills, avoidance of negativity, social support, pro-social behavior, positive psychology, and spiritual connectedness, along with proper sleep, exercise, and diet are your keys to developing a strong integrative health index.

Oxidative stress and NF-kB

Mechanisms involving oxidative stress, and one of its key mediators, the pro-inflammatory gene transcription factor NF-kB (responsible for producing pro-inflammatory cytokines), explain how stress is implicated in the genesis and progression of chronic diseases according to the mind–body integrative medicine equation. The statistical association between stress and these disorders is clear, and although more research is required to unequivocally establish a causal link, consistent evidence suggests that this causal relation is real. NF-kB can be activated by psychosocial stress and may then directly target the functioning of the cells that line our blood vessels. Thus, it is an additional risk factor for cardiovascular, cerebrovascular, and renal disease. Components of the metabolic syndrome, such as hypertension, obesity, insulin-resistant diabetes mellitus, and high cholesterol can also activate NF-kB, as can psychologically produced oxidative stress, cytokines, growth factors, and other stress-affected molecules.

NEW RESEARCH ON NF-kB

Fascinating new research, some of it done at the Benson-Henry Institute (BHI) for Mind Body Medicine and Integrative Health at Massachusetts General Hospital (US), brings these findings down to the cellular level. Several mind–body interventions that elicit the relaxation response, including meditation and yoga have been found to reduce the activation of the NF-kB gene pathway and also to bolster oxidative metabolic efficiency. It may be that gene expression changes, induced environmentally in an epigenetic fashion, may form the links between your personal genetic endowment and resilience, and your genetic vulnerability and stress.

In Sweden, a large public health study followed 237,980 men who were being inducted into the Swedish Army in the years 1952–56. During a follow-up period between 1987 and 2010, it was found that while higher physical fitness lowered the risk of heart disease, this effect was significantly reduced in those with lower stress resilience. In other words, low resilience in the face of stress will lower the benefit of exercise when it comes to the risk of heart disease. In a reflection of the mind–body medicine equation, the likelihood of healthy outcomes is endangered in the setting of low stress resilience.

How might we understand this relationship biologically at the cellular level? In keeping with the discussion above, it may be that the illness vectors described in these studies are effecting changes in cellular gene expression leading to disease-provoking protein profiles.

Meeting the stress challenge head on

Insecure attachment style as a residual developmental trait, often in response to adverse childhood events and toxic stress in childhood, may elevate one's illness index because it creates negative threats as opposed to positive appraisals. Subsequent relationship stress thus reduces resilience. Evidence suggests that attachment insecurity and risk for disease are associated. This risk is mediated by susceptibility to stress and allostatic loading as well as maladaptive attempts to increase resilience through misuse of external mediators (for example, drugs and alcohol), and inappropriate use of protective factors (for example, medical overutilization). The problem is compounded once illness takes hold, because illness itself is a stressor that ignites the threat of separation. This is all the more reason that societies worldwide need to redouble their efforts to reduce childhood toxic stress. There is no other single effort that would in the long run provide greater impact in terms of our health promotion and illness prevention.

The medical profession should take heed of this equation when they are evaluating patients. This is because during the clinical encounter it is important to get a more objective impression of the patient's levels of stress and resilience. This helps the clinician to understand the patient's separation challenges and attachment status, and what can be done to improve the ratio, thereby increasing the integrative health index.

Mind–body integrative medicine interventions provide ways for you to reduce your stress. These include meditation, building resilience through social support and pro-social behaviors, improving cognitive skills, positive psychological approaches, and spiritual connectedness. Along with healthy behaviors this will promote in you a secure attachment style and an enriched environment of self-care and social connection. This new approach can be effective in buffering against stress and in building your resilience. And this approach may give all of us the best chance we have to remain healthy in a challenging world.

SPIRITUALITY

If the sense of attachment between patient and doctor can access the healing power of optimism, then it is likely that connectedness with a believed-in power, or source of love greater than us, will also be associated with health benefits. Spiritual connectedness has the power to increase positive expectations, while at the same time strengthening social support. Spirituality also often encompasses a meditative tradition that aids the relaxation response. The spiritual life in this way can often be successful in building the foundations for our resilience. *The Buddha* by Odilon Redon, c. 1905.

GLOSSARY

ACUTE An adjective meaning sudden or short-lived.

ADRENAL CORTEX The outermost layer of the adrenal glands.

ADRENAL GLANDS Two organs, located just above the kidneys, which produce a range of hormones, several of which are involved in the body's response to stress.

ADRENAL MEDULLA The inner part of the adrenal gland.

ADRENALINE An alternative name for epinephrine.

ADRENOCORTICOTROPIC HORMONE (ACTH) A hormone released by the pituitary gland as part of the body's response to stress.

ALLOSTASIS The process by which the body changes in order to maintain stability in the face of physical or mental stressors.

ALLOSTATIC LOAD The effects of allostasis from chronic stress, which manifest as wear and tear on tissues or inside cells, and non-communicable diseases.

AMYGDALA A region of the brain in the limbic system heavily involved in emotional responses, particularly fear, and in processing memories. There are two amygdalae, one in each brain hemisphere, below the cortex, very close to the hippocampus.

ANTERIOR CINGULATE CORTEX The front part of the cingulate cortex of the brain, just beneath the neocortex. It processes emotional signals from the amygdala, as well as general sensory information from the thalamus, and is involved in attachment, mediation of pain, and decision-making under stress.

ANTIGEN Any molecule that the immune system identifies as belonging to something foreign and potentially harmful, such as a virus.

ATHEROSCLEROSIS Also known as "hardening of the arteries." A disease in which a plaque of cholesterol, fat, and calcium builds up on the endothelium of arteries.

AUTONOMIC NERVOUS SYSTEM (ANS) Part of the peripheral nervous system that controls organs and modulates other bodily functions. Nerve signals in the ANS are not normally under conscious control.

B CELL A type of white blood cell that matures in the bone marrow (hence the "B") and is a crucial part of the acquired immune system.

BRAIN-DERIVED NEUROTROPHIC FACTOR (BDNF) A protein made in the brain that encourages the proliferation of new neurons and synapses, and protects existing ones.

BRAIN STEM An evolutionarily ancient part of the brain situated at the top of the spinal cord. All signals passing between brain and body pass through it.

CARDIOVASCULAR DISEASE Any disease involving the heart (*cardio*) or the blood vessels (*vascular*), for example stroke, angina, and heart attack.

CATECHOLAMINE Any compound whose molecules have an amine group (NH_2) and a catechol group (a carbon ring with two hydroxyl groups, OH), for example epinephrine, norepinephrine, and dopamine.

CEREBRAL CORTEX See **CEREBRUM** and **CORTEX**.

CEREBRUM The uppermost and largest part of the brain, whose cortex, deeply folded with gyri and sulci, is the most striking and recognizable feature of the brain. The cerebrum is divided into two hemispheres, and has areas that map to the senses and that initiate control of muscles, as well as areas that process language and "executive control."

CHROMOSOME Any of forty-six separate strands of DNA inside each cell. At each end of a chromosome is a region called a telomere, which sacrificially helps protect against damage to the DNA.

CHRONIC An adjective meaning "persisting over a long period of time."

CINGULATE CORTEX Part of the cerebrum, located on the inner surface of each cerebral hemisphere. It is the part of the paralimbic cortex that lies between the cingulum bundle and the limbic system involved in emotional processing, and the neocortex involved in cognition.

COGNITIVE BEHAVIOR THERAPY (CBT) A treatment for psychological conditions such as depression, anxiety, and post-traumatic stress disorder.

CORTEX The outermost layer of an organ. Most notable are the cortices of the cerebrum and the adrenal glands.

CORTICOTROPIN-RELEASING HORMONE (CRH) A hormone released by the hypothalamus as part of the body's response to stress. CRH stimulates the pituitary gland to produce another hormone, as part of the HPA axis.

CORTISOL A glucocorticoid steroid hormone produced by the adrenal cortex in the last phase of the HPA axis, part of the body's response to stress. It prepares the body to be more alert and ready to face challenges, by raising blood sugar concentration and diverting energy from essential activities such as immune system function and bone growth.

CYTOKINE A protein chemical messenger, produced and released by one cell to produce a change in another. Some cytokines are key players in the immune system—in particular, the inflammatory response—and in wound healing.

DISTRESS A term used to describe a state when a patient is unable to adapt to or deal with stressors.

DNA Deoxyribonucleic acid. The chemical compound whose molecules encode genes, "spelled out" by variations in the molecule.

DORSOLATERAL PREFRONTAL CORTEX Part of the prefrontal cortex of the brain involved in executive functions, for example decision-making, and the planning and inhibition of movements.

ENDOGENOUS OPIOIDS Opioids that are produced inside the body. The best-known examples are endorphins.

ENDOTHELIUM The inner lining of blood vessels.

EPIGENETICS The study of the factors that affect the expression of genes.

EPINEPHRINE Also called adrenaline. One of the key hormones involved in the body's fight-or-flight response to stress, epinephrine is released by the adrenal medulla under control of the sympathetic nervous system.

EUSTRESS A term used to describe the normal, short-lived stressors of everyday life to which the body can normally adapt, by allostasis.

FIBRINOGEN A protein that encourages the formation of blood clots.

FIGHT-OR-FLIGHT RESPONSE The sympathetic nervous system's main reaction to stressful situations, in which a series of hormone releases cause the brain to become more alert and prepare the body for action, by increasing heart rate, respiratory rate, and blood sugar.

GABA Gamma-aminobutyric acid, a neurotransmitter that acts at synapses between neurons to inhibit activity. Its presence in the brain promotes relaxation and reduces the stress response and anxiety, and levels increase by practicing meditation and relaxation.

GENE A portion of DNA that carries instructions for building a specific protein. The human genome consists of around 20,000 genes.

GLUCOCORTICOID A class of steroid hormones, of which cortisol is the most notable.

GLUCOSE A sugar compound produced by the breakdown of carbohydrates in food. Glucose circulates in the blood, providing fuel for cellular activity.

GLYCOGEN An energy storage compound produced from glucose in muscle and liver cells. Insulin encourages the production of glycogen when blood sugar (glucose) levels are high. Cortisol and epinephrine encourage the breakdown of glycogen, to form available glucose for a ready supply of energy, as part of the body's response to stress.

GYRUS (plural gyri) Any "peak" in the wavy, heavily folded structure of the cerebral cortex, useful in brain anatomy.

HIPPOCAMPUS (plural hippocampi) Part of the limbic system of the brain, involved in the formation of memories, dampening of amygdalar tone as well as spatial awareness. There is one hippocampus in each cerebral hemisphere.

HOMEOSTASIS Any process in the body that maintains a particular state, such as the concentration of oxygen in the blood, body temperature, and blood pressure.

HORMONE A compound produced by glands that circulates in the blood, effecting changes in organs. The human body produces around seventy different hormones, including epinephrine, norepinephrine, insulin, and cortisol.

HPA AXIS Abbreviation for the hypothalamic-pituitary-adrenal axis. A key part of the body's stress response, which begins with the hypothalamus secreting corticotropin-releasing hormone, which stimulates the pituitary gland to release adrenocorticotropic hormone (ACTH). At the adrenal glands, ACTH stimulates the adrenal cortex to release cortisol, which encourages the adrenal medulla to secrete epinephrine and norepinephrine.

HYPERTENSION The scientific term for high (hyper-) blood pressure (tension). Hypertension is a major factor in cardiovascular disease.

HYPOTHALAMUS A small region of the brain, below the cerebrum and close to the brain stem, that produces a range of hormones that govern variables such as body temperature, hunger, thirst, and sleep. As part of the HPA axis, its secretion of corticotropin-releasing hormone plays a key role in the body's response to stress.

IMMUNE SYSTEM The collection of cells, tissues, and processes that defends the body against disease, in particular from pathogens such as bacteria and viruses.

INFLAMMATION A process in the innate, or generic immune system that produces a cascade of chemical reactions resulting in the recruitment of macrophages and the production of reactive oxygen species. Acute inflammation kills pathogens and encourages healing, but chronic inflammation is a major factor in cardiovascular disease and other non-communicable diseases.

INFLAMMATORY RESPONSE Acute inflammation.

INSULAR CORTEX Part of the paralimbic cortex of the brain that is involved in many visceral processes, including emotional experience and brain–heart connection.

INSULIN A hormone produced by specialized cells in the pancreas that is released when blood glucose concentration is raised. It has the effect of increasing storage of glucose, as glycogen.

LIMBIC SYSTEM A collection of structures in the middle, subcortical regions of the brain that provide emotional tone to sensory experiences. These structures include hippocampus, amygdala, fornix, septum, mammillary bodies and cingulum.

LOCUS COERULEUS Part of the brain that is heavily involved in the body's response to stress. It is the brain's main producer of norepinephrine, which it releases at synapses in many other parts of the brain, including the amygdala.

MACROPHAGE A type of white blood cell that is the main component of the innate immune system.

MEDIAL PREFRONTAL CORTEX Part of the prefrontal cortex that is medial, i.e. in the middle of each cerebral hemisphere. It helps to reduce amygdala excitement during threat.

METABOLIC SYNDROME A combination of high lipids, high blood pressure, truncal obesity, and poor insulin receptor functioning that leads to type 2 diabetes due to insulin resistance. It can be caused by unhealthy diets, but also by chronic stress. Metabolic syndrome greatly increases the risk of cardiovascular disease and other stress-related, non-communicable diseases.

MICROBIOTA The enormous collection of beneficial, benign, or pathogenic microbes that live inside the human body, primarily in the gut.

MITOCHONDRIA (singular mitochondrion) Organelles found in every cell that play a key role in cellular respiration, the process that extracts useful energy from glucose.

MONOAMINES A class of neurotransmitters that possess a single amine group (NH_2), for example serotonin and histamine. The most important monoamines in the body's response to stress are the catecholamines, epinephrine and norepinephrine.

NEOCORTEX/NEOCORTICAL AREAS The main part of the cerebral cortex of the brain, in which most of the higher cognitive functions seem to take place.

NEUROCHEMICAL Any compound that specifically takes part in the activity of neurons.

NEUROGENESIS The generation of new neurons (nerve cells).

NEUROPEPTIDE A protein neurochemical used as a signaling molecule by neurons.

NEUROTRANSMITTER A type of neurochemical that is produced inside neurons and released at synapses.

NF-kB Nuclear factor kappa-light-chain-enhancer of activated B cells, a protein involved in the inflammatory response. Chronically raised levels of NF-kB lead to chronic inflammation, which increases the risk of non-communicable diseases.

NON-COMMUNICABLE DISEASE (NCD) Any disease that is not produced by transmission from one person to another of a pathogen, such as a bacterium or virus, and is therefore not infectious.

NORADRENALINE See **NOREPINEPHRINE**.

NOREPINEPHRINE One of the key chemical compounds involved in the body's response to stress. It acts as both a neurotransmitter (at synapses) and a hormone (in the bloodstream).

NUCLEUS ACCUMBENS A small region of the brain close to the hypothalamus. The nucleus accumbens plays important roles in reward, pleasure, and addiction.

OPIOIDS Compounds that have the same effect as morphine, i.e. that reduce pain. Manufactured opioids are used in drugs, but the body produces its own endogenous opioids such as endorphins.

ORBITOFRONTAL CORTEX Part of the prefrontal cortex of each cerebral hemisphere, just above and behind the eyes. Experiments suggest that it is involved in high-level decision-making.

OXIDATIVE STRESS Damage caused to molecules inside a cell, most notably DNA, by reactive oxygen species. Oxidative stress is enhanced by chronic stress, and is implicated in many non-communicable diseases, such as cancer and cardiovascular disease.

PARALIMBIC CORTEX Part of the cerebral cortex (though not the neocortex) in the brain that communicates with structures that make up the limbic system, in particular the amygdala.

PARASYMPATHETIC NERVOUS SYSTEM Part of the autonomic nervous system. Its function is to return the body to normal after a fight-or-flight stress response.

PERIPHERAL NERVOUS SYSTEM That part of the nervous system that lies outside the brain and the spinal cord.

PIRIFORM CORTEX Part of the brain involved in the sense of smell that communicates with the limbic system, in particular the amygdalae.

PITUITARY GLAND A gland (hormone-producing organ) in the brain, located just above the brain stem, that secretes several important hormones, notably adrenocorticotropic hormone, which is involved in the HPA axis.

POST-TRAUMATIC STRESS DISORDER (PTSD) A psychological condition caused by a person experiencing a traumatic event. Symptoms include troubled sleep, avoiding reminders of the event, anxiety, and irritability or aggression.

PRIMARY SENSORY CORTEX More correctly, the primary somatosensory cortex. Part of the cerebral cortex where nerve signals from touch receptors around the body arrive. It is located in one gyrus in each of the cerebral hemispheres.

RAPHE NUCLEI Part of the reticular formation, a region of the brain stem that produces the neurotransmitter serotonin, which regulates mood.

REACTIVE OXYGEN SPECIES (ROS) Highly chemically active molecules that can cause damage to other molecules, particularly DNA. They are a normal bi-product of metabolism, the set of chemical reactions inside cells, and mechanisms exist within cells to eradicate them or mend the damage they cause.

RETICULAR FORMATION Part of the brain stem that is involved in consciousness, breathing, and producing the neurotransmitter serotonin.

SENSORY CORTEX Any part of the cerebral cortex where signals from sensory neurons arrive for processing.

SEPTAL REGION Also called the septum or septal area. A region near the center of the brain that has good connections to the hippocampus and the hypothalamus. It is part of the limbic system.

STEROID Any compound whose molecules have a particular arrangement of carbon rings—the steroid configuration.

STRESS The way the body responds to stressors.

STRESSOR Anything that could threaten an organism's well-being or survival. Stressors can be psychological (mental) or physical.

SULCUS (plural sulci) Any trough in the wavy, heavily folded cerebral cortex.

SYMPATHETIC NERVOUS SYSTEM (SNS) Part of the autonomic nervous system. Its function is to instigate the fight-or-flight stress response, dilating the pupils, increasing heart rate, reducing the rate of digestion, and inhibiting urination.

SYNAPSE A small gap between two neurons that allows one neuron to influence the behavior of the other.

T CELL A type of white blood cell that matures in the thymus (hence the "T") and is a crucial part of the acquired immune system.

TELOMERES Regions at the ends of chromosomes that are sacrificially degraded each time a chromosome is replicated, to protect the DNA in the bulk of the chromosome, which carries instructions for proteins.

TEMPORAL LOBE A major part of each cerebral hemisphere, located just behind the ears. The temporal lobes are home to visual memory, several features of language, and some emotional processing.

THALAMUS A region of the brain below the cerebral cortex, at the very center of the brain. The thalamus receives all the inputs from sensory neurons (except those from the nose) and relays them on to various parts of the brain. It also seems to play a part in sleep and arousal.

TOXIC STRESS A term used to describe abuse or neglect in childhood that often leads to chronic, mental health issues later in life, or even non-communicable diseases (NCDs).

VENTRAL TEGMENTAL AREA Sits deep inside the brain, close to the hippocampus, which is heavily involved in feelings of addiction, pleasure, and reward. Its neurons connect with many other parts of the brain. It appears to play a part in motivation.

SELECTED BIBLIOGRAPHY

ASTIN, J. A., and S. L. SHAPIRO, D. M. EISENBERG, K. L. FORYS. "Mind–body medicine: state of the science, implications for practice." *The Journal of the American Board of Family Practice.* (2003) 16: 131–47.

BARROWS, K. A., and B. P. JACOBS. "Mind–body medicine. An introduction and review of the literature." *Medical Clinics of North America.* (2002) 86: 11–31.

BHASIN, M. K., and J. A. DUSEK, B. H. CHANG, M. G. JOSEPH, J. W. DENNINGER, G. L. FRICCHIONE, H. BENSON, T. A. LIBERMANN. "Relaxation response induces temporal transcriptome changes in energy metabolism, insulin secretion and inflammatory pathways." *PLoS One.* (2013) 8: e62817.

BOWLBY, J. *Attachment and Loss*, 2nd edition. New York: Basic Books, 1982.

EPEL, E. S., and E. H. BLACKBURN, J. LIN, F. S. DHABHAR, N. E. ADLER, J. D. MORROW, R. M. CAWTHON. "Accelerated telomere shortening in response to life stress." Proceedings of the National Academy of Science of the USA. (December 7, 2004) 101(49): 17312–5.

FEDER, A., and E. J. NESTLER, D. S. CHARNEY. "Psychobiology and molecular genetics of resilience." *Nature Reviews Neuroscience.* (2009) 10: 446–57.

FRICCHIONE, G. L. *Compassion and Healing in Medicine and Society. On the Nature and Uses of Attachment Solutions to Separation Challenges.* Baltimore: Johns Hopkins University Press, 2011.

LAZAR, S. W., and G. BUSH, R. L. GOLLUB, G. L. FRICCHIONE, G. KHALSA, H. BENSON. "Functional brain mapping of the relaxation response and meditation." *Neuroreport.* (2000) 11: 1581–5.

McEWEN, B. S., and P. J. GIANAROS. "Central role of the brain in stress and adaptation: links to socioeconomic status, health, and disease." *Annals of the New York Academy of Sciences.* (February 2010) 1186: 190–222.

McEWEN, B. S. "Protective and damaging effects of stress mediators." *The New England Journal of Medicine.* (1998) 338: 171–9.

MILLER, G. E., and E. CHEN, J. SZE, T. MARIN, J. M. AREVALO, R. DOLL, R. MA, S. W. COLE. "Functional genomic fingerprint of chronic stress in humans: blunted glucocorticoid and increased NF-kappaB signaling." *Biological Psychiatry.* (2008) 64: 266–72.

NABI, H., et al. "Increased risk of coronary heart disease among individuals reporting adverse impact of stress on their health: the Whitehall II prospective cohort study." *European Heart Journal.* (2013) 34(34): 2697–2705.

NAGAI, M., and S. HOSHIDE, K. KARIO. "The insular cortex and cardiovascular system: a new insight into the brain–heart axis." *Journal of the American Society of Hypertension.* (July–August 2010) 4(4): 174–82.

NARAYAN, K. M., and M. K. ALI, J. P. KOPLAN. "Global noncommunicable diseases—where worlds meet." *The New England Journal of Medicine.* (2010) 363: 1196–8.

PROVENCAL, N., and E. B. BINDER. "The effects of early life stress on the epigenome: from the womb to adulthood and even before." *Experimental Neurology.* (June 2015) 268: 10–20.

ROOZENDAAL, B., and B. S. McEWEN, S. CHATTARJI. "Stress, memory and the amygdala." *Nature Reviews Neuroscience.* (2009) 10: 423–33.

RUBERMAN, W., and E. WEINBLATT, J. D. GOLDBERG, B. S. CHAUDHARY. http://www.ncbi.nlm.nih.gov/pubmed/6749228?ordinalpos=1&itool=EntrezSystem2.PEntrez.Pubmed.Pubmed_ResultsPanel.Pubmed_Default ReportPanel.Pubmed_RVDocSum" "Psychosocial influences on mortality after myocardial infarction." *The New England Journal of Medicine.* (1984) 311: 552–9.

SAMUELSON, M., and M. FORET, M. BAIM, J. LERNER, G. L. FRICCHIONE, H. BENSON, J. DUSEK, A. YEUNG. "Exploring the effectiveness of a comprehensive mind–body intervention for medical symptom relief." *Journal of Alternative and Complementary Medicine.* (2010) 16: 187–92.

STAHL, J. E., et al. "Relaxation response and resiliency training and its effect on healthcare resource utilization." *PLoS ONE.* (2015) 10(10): e0140212.

SHONKOFF, J. P. "Leveraging the biology of adversity to address the roots of disparities in health and development." Proceedings of the National Academy of Science of the USA. (October 16, 2012) 109, Supplement 2: 17302–7.

SOUFER, R., and H. JAIN, A. J. YOON. "Heart-brain interactions in mental stress-induced myocardial ischemia." *Current Cardiology Reports.* (2009) 11(2): 133–40.

STEPTOE, A., et al. "Disruption of multisystem responses to stress in type 2 diabetes: investigating the dynamics of allostatic load." Proceedings of the National Academy of Science of the USA. (November 4, 2014) 111(44): 15693–8.

STEPTOE, A., and A. J. MOLLOY. "Personality and heart disease." *Heart.* (July 2007) 93(7): 783–4.

World Health Statistics 2010 report http://www.who.int/whr/en/index.html.

INDEX

ACKNOWLEDGMENTS

AUTHOR ACKNOWLEDGMENTS

The authors together would like to thank the Benson-Henry Institute for Mind Body Medicine at Massachusetts General Hospital (MGH) staff and associates, in particular Dr. Herbert Benson (Director Emeritus). We also express our appreciation to Dr. Jerrold Rosenbaum, our MGH Department of Psychiatry Chair, and Dr. Maurizio Fava, our Departmental Vice Chair and MGH Director of Clinical Research, for their support and mentorship.

We also want to recognize the leaders in stress research who have spoken at our annual meetings and at our Institute for the wisdom they imparted, much of which has found its way into this book. These include Dr. Jon Kabat-Zinn, Dr. Bruce McEwen, Dr. Sonia Lupien, Dr. Anthony Biglan, Dr. Margaret Chesney, Dr. Steven Southwick and Dr. Mohamed Milad.

I'd personally like to thank my wife Kathryn and my children, Kristen, Marielle and Jon and my son-in-law Erich, for their love and support, without which the stress in my life would undoubtedly morph into distress!
GREGORY L. FRICCHIONE, MD

I wish to thank my wife, Sharon, and our two daughters, Janet and Alicia, for their support, and Run Feng, Max Martinson, and Alicia Yeung for their assistance.
ALBERT YEUNG, MD, ScD.

I'd like to thank my husband Sean, without whose support I wouldn't have been able to dedicate hours at the library toward this project, and my 2-year-old daughter Sofia, who along with her father has taught me more about stress-buffering, feel-good chemicals than I could ever learn from a book. I also want to thank my co-author, Gregory Fricchione, for inviting me to join this project and recognizing my passion for the subject matter of this book. Lastly, I'd like to thank my parents for their constant support and inspiration.
ANA IVKOVIC, MD